Lecture Notes in Biomathematics

Managing Editor: S. Levin

25

Peter Yodzis

Competition for Space and the Structure of Ecological Communities

Springer-Verlag
Berlin Heidelberg New York 1978

Author

Peter Yodzis
Institut für Theoretische Physik
Universität Bern
Sidlerstraße 5
3012 Bern/Switzerland

Library of Congress Cataloging in Publication Data

Yodzis, Peter, 1943-
  Competition for space and the structure of ecological
communities.

  (Lectures notes in biomathematics ; v. 25)
  Bibliography: p.
  Includes indexes.
  1. Competition in ecology--Mathematics. 2. Biotic
communities--Mathematics. I. Title. II. Series.
QH546.3.Y63   574.5'24   78-11398

AMS Subject Classifications (1970): 34A45, 34C05, 34D10, 65C05, 92A15

ISBN-13: 978-3-540-08936-0   e-ISBN-13: 978-3-642-93097-3
DOI: 10.1007/978-3-642-93097-3

PREFACE

This volume is an investigation of interspecific competition for
space, particularly among sessile organisms, both plant and animal,
and its consequences for community structure. While my own contribution —— and the bulk of this volume —— lies in mathematical analysis
of the phenomenon, I have also tried to summarize the most important
natural historical aspects of these communities, and have devoted much
effort to relating the mathematical results to observations of the
natural world.

Thus, the volume has both a synthetic and an analytic aspect.
On the one hand, I have been struck by certain similarities among
many communities, from forests to mussel beds, in which spatial competition is important. On the other hand, I have analyzed this phenomenon by means of reaction-dispersal models. Finally, the mathematical
analysis has suggested a conceptual framework for these communities
which, I believe, further unifies and illuminates the field data.

A focal perception of this work is that, just as niche relations
provide an appropriate expression of the influence of resource competition on community structure, so do dominance relations provide an
appropriate expression of the influence of spatial competition.

I would hope that people in the entire spectrum of those with an
interest in ecology, from field biologists seeking a conceptual framework for their observations to mathematicians seeking fruitful areas
of application, could find stimulation here. I fear, however, that
every reader will find some parts of the text trivial and some parts
of it incomprehensible —— with these parts different for any two
readers. In view of this, the bibliography, while far from comprehen-
sive, does attempt to provide an entry to the literature starting from
almost any point touched upon here. In addition, there is a subject

index and, for those readers who are already familiar with the literature, an author index.

*

This work could not have been done without the broad-mindedness
and encouragement of my colleagues at the Institute for Theoretical
Physics of Bern University, especially Petr Hájíček, Heinrich Leutwyler,
and André Mercier. The assistance of Paddy Maloney and his colleagues
in the labors of writing and rewriting, and of my wife Karin in the
preparation of the final manuscript, was indispensible.

This volume has benefited from discussions with Menk Amacher,
Joseph Connell, Jared Diamond, Hans-Rudolf Gnägi, Alan Held, Simon
Levin, Robert May, Jon Roughgarden, Jakob Schmid, and Karin Yodzis.
Correspondence with Paul Dayton and Robert Paine was also most stimu-
lating. Ted Case, Hal Caswell, Edward Conway, Simon Levin, Jane Lubchenco Menge, James Porter, and Joel Smoller very kindly sent me manu-
scripts of some of their work prior to publication. I am especially
grateful to Paul Dayton, Alan Held, Simon Levin, Robert May, J. J.
Oertli, Robert Paine, and Karin Yodzis for taking the time and energy
to read and comment on earlier versions of this text.

The result is vastly better than it would have been without the
generous assistance of these people. The blame for any remaining
inconsistencies, inaccuracies, imprecisions, omissions, or errors is
entirely mine.

- Peter Yodzis

CONTENTS

Chapter I.   INTRODUCTION

I.1. Competition for space.

   The requisites for life are, in any habitat, limited in supply.
Therefore, the survival of each species of life depends upon its ability
to compete with other species for these requisites. This simple observation has become a major unifying theme in contemporary theoretical
ecology.

   In the case of competition for renewable resources (for instance,
food), we now have a sophisticated and highly developed theory, exem-
plified by Robert MacArthur's book (1972). Other varieties of competition have been less thoroughly studied; among these, competition for
space (Lebensraum) is likely to be especially important for community
structure.

   Strictly speaking, every organism must have space in which to live,
so that at sufficiently high population densities space will always
become a limiting factor. However, for many organisms other factors
will already be limiting at much lower densities: for these species
competition for space will under realistic conditions play only a very
minor role. Our interest in this study will be rather in organisms for
which space is, already at densities which normally occur in nature,
a limiting or at least a very important factor.

   We may expect, for example, that primary producers (green plants
and algae) will often be space-limited, because they need, by definition, sunlight in order to live, and this can be thought of as distributed more or less uniformly over the surface of the Earth (but see the
remarks below on stratification). In the animal kingdom, the most obvious examples of likely space-limitation are provided by the sessile
species (creatures which spend most of their lives attached to some
substratum). Let us examine some of the specific mechanisms by which

competition for space is carried out among these organisms.

Plants have a physiological need for light, carbon dioxide, water, and nutrients. In order to obtain light, a plant must appropriate a certain amount of space on the Earth's surface; in order to obtain water and nutrients, a terrestrial plant must appropriate a certain amount of (three-dimensional) space below ground level, whence, again, a certain amount of space on the Earth's surface. For example, the scarcity of water in desert regions dictates a widely-spaced pattern of shrubs there (Beals 1968; Yeaton, Thompson, and Cody, cited by Cody 1974).

Competition for light is mediated largely by morphology above ground level, competition for water and nutrients by morphology below ground level. Thus one factor which determines the forms of leaves and their distributions on plants is the extent to which each species invests energy in competition for light (for instance: Kan-Ichi Sakai 1961; Horn 1971; Grime 1977). It is well known that at higher plant densities root development is inhibited (e.g., Dittmer 1937, 1938), and root morphology has been directly implicated in competition for water and nutrients (Moore 1959; McCowan and Williams 1968; Sims and Mueller-Dombois 1968; Cable 1969; Harris and Wilson 1970).

These two aspects of competition for space among plants are neatly brought out in studies by Evans (1960) of competition between the broad-leafed plant Erodium botrys and two (more narrow-leafed) grass species. When nitrogen levels in the soil are high, Erodium suppresses the grasses; at lower nitrogen levels the grasses are more successful. Both results can be understood in terms of morphology and competition for space: with plentiful nitrogen the competition is mainly for light and the shade-producing morphology of Erodium is a decisive advantage, but when nitrogen is scarce this advantage of Erodium is offset by the superior ability of the grasses to obtain it through their extensive root systems.

Some readers may find this emphasis on space peculiar: why not just say that light is limiting, or nitrogen? The point is that in order to win light or nitrogen, a plant must win space: the dynamic of plant populations must be a spatial dynamic. If we sometimes refer to plant populations as space-limited, this is not to deny that for some purposes they are more fruitfully thought of as light-limited, or nitrogen-limited, for example, but to acknowledge the central role played by the spatial aspect of plant requirements in plant population dynamics.

Returning to our discussion of the specific mechanisms of competition among plant species: another quite common such mechanism is allelopathy: the production by one species of chemicals which are toxic to other species (see, for instance, the review articles by Woods 1960 and by Whittaker and Feeny 1971). Since these chemicals are generally present only in the immediate neighborhood of the plant that produces them, they have the effect of keeping susceptible species at a certain distance, so that, again, this amounts to competition for space. Indeed, in a classic example, the invasion of grassland by chaparral shrubs, the shrub patches are surrounded by belts of bare soil 1 to 2 meters wide, due at least in part to herbicides produced by the shrubs. (Some workers, such as Bartholomew (1970), maintain that vertebrate grazers play an essential role in maintaining these bare zones, but others, for instance Muller and del Moral (1971), reject this interpretation.)

While it is clear that allelochemicals often do act as agents for interspecific competition, it is by no means clear that production of these chemicals originally evolved because of interspecific competitive pressure; indeed, sometimes these chemicals are autotoxic as well (see Whittaker and Feeny 1971 for examples). Among alternative suggestions put forward for the evolutionary origin of allelopathy are population self-regulation (Went 1970) and plant metabolism (Seigler and Price 1976).

A very similar mechanism enables the succelent herb Mesembryanthemum crystallinum to invade grasslands along the central coast of California, USA (Vivrette and Muller 1977). This species accumulates salt throughout its lifetime; when a plant dies the salt leaches out into the soil, creating an osmotic environment which other species in the system cannot tolerate. Vivrette and Muller refer to this effect as osmotic interference.

An unusual but very beautiful example of spatial competition among plant species is provided by mutualistic associations between certain trees (myrmecophytes) and ants which inhabit them, such as the Acacia-Pseudomyrmex and Cecropia-Azteca associations (Janzen 1966, 1967, 1969). The trees are highly adapted to the ants, even to the point of producing substances which are rather clearly of no direct benefit to the plants themselves, but are the primary food source of the ant colony. In

return, the ants, which are highly antagonistic to plants which do not
belong to their host species, win space for the host plants: they chew
off foliage from other species which comes into contact with their host
plant, with occupied plants in some cases virtually eating their way
through the overtopping canopy. Pseudomyrmex will even kill vegetation
at distaces up to 2 meters from a host Acacia plant. This is literally
competition for space in which to grow.

The spatial requirements of sessile fauna are rather simpler than
those of plants: a sessile animal needs simply a certain area of substratum (stream bed, rock, etc.) to which to attach. Competition among
these creatures takes also for the most part a simple form: direct physical aggression. In a classic study of competition, Connell (1961a)
found that one species of barnacle outcompeted another for attachment
space by means of undercutting or even smashing of individuals. Similar
behavior has been observed on the part of mussels (Dayton 1971).
Sponges overgrow competitors (Rützler 1965; Sarà 1970), and at least one
Antarctic species, Mycale acerata, appears to digest the overgrown competitors (P. K. Dayton, private communication), as do many overgrowing
forms of coral (Lang 1972, 1973). Allelopathic effects, similar to
those just described for plants, sometimes also play a role in competition among coral reef invertebrates (Jackson and Buss 1975).

J. B. C. Jackson (1977a) has made a compilation, reproduced here
in table 1, of competitive mechanisms among the major groups of marine
epifaunal metazoa that attach permanently to hard substrata. The competitive mechanisms (listed in the seventh column of table 1) almost
always involve direct aggression for attachment space.

It is, then, clear that competition for space is of major importance for sessile organisms, both plant and animal, and that space must
often be considered limiting for these organisms. This study will be
concerned almost entirely with sessile organisms, but this is not to
suggest that these are the only creatures which compete strongly for
space. For instance, sympatric species which display interspecific
territoriality should probably also be thought of as competing for
space. Some workers (Sale 1972; Smith and Tyler 1972) have suggested
that suitable living space may actually be a limiting factor for many
(territorial) coral-reef fishes, and Sale (1975) describes the behavior
of a guild of these fishes in terms which strongly suggest competition
for patches of refuge space on the reef. It could also prove fruitful

to apply these ideas to the bird guilds studied by Diamond (1975) in
tropical archipelagos.

*

We shall in the remainder of this section discuss some of the uses
to which one would hope to be able to put a theory of competition for
space. In the next section we discuss niche relations for communities
of sessile organisms, and in Section 3 we formulate mathematical models
for studying spatial competition. In Sections 4 and 5 we characterize
the particular approach to these models that we shall take here, and
in Section 6 we sketch the contents of the remainder of the volume.

*

Our interest throughout this study will be in the implications of
spatial competition for community structure: thus we will treat assemblages of (possibly many) species competing for space. In the case of
sessile organisms, some characteristic community properties are:

1. <u>Patchiness</u>. The communities often have the appearance of a
patchwork of different species, rather than a spatially uniform mixture
(Drew 1942; Ashby 1948; Potzger and Potzger 1950; Anderson, Loucks, and
Swain 1969; Pielou 1969; Dayton 1973a; Kershaw 1973; Richards and Williamson 1975; Williamson 1975; Whittaker and Levin 1977). There is a
tendency for rare species to be more "clumped" spatially than common
ones (Hairston 1959; Kershaw 1963; Lloyd 1967).

2. <u>Numerical dominance</u>. Often one or a few species are far more
abundant than any others; definite patterns tend to be seen in the
relative abundances of the species (Whittaker 1965; May 1975; Pielou
1975; and the further references cited in Chapter III).

3. <u>Succession</u>. As the community develops from the colonization
of bare space to some sort of steady state, there are often characteristic developments in various community properties, such as biomass,
species richness, and equitability (Kershaw 1973; Horn 1974; and the
additional references cited in Section III.5).

A theory of competition for space should explain these regulari-
ties. In addition, there are a number of general questions about community structure which such a theory must elucidate for communities
where spatial competition is important. For instance:

4. <u>Species packing</u>. How many species can be "packed" into a given

TABLE I.1

Ecological and Life History Characteristics of Solitary and Colonial Attached Invertebrates

Invertebrate	Morphology	Growth Pattern	Method of Growth	Asexual Reproduction	Sexual Reproduction	Competition Mechanisms	Susceptibility to Epizoan Recruitment	Defense against Predation
Solitary: Serpulidae and Sabellariidae.	Encrusting; erect; or both when "solitary," massive "colonies" erect and/or intertwined individuals	Determinate; exponentially decreases with age	Simple	Usually absent (transverse fission of *Filograna*)	May rapidly attain sexual maturity (3–6 wk); many brood eggs; also produce larvae; occasional self-fertilization	Overgrowth; some grow in erect forms when faced with overgrowth; "colony" surfaces all apertures (not overgrown)	High on tubes except for "colony" surfaces of tube apertures	Tube
Bivalvia	Encrusting; erect	Determinate; usually exponentially decreases with age	Simple	Absent	Maturity at various ages (more often years than months); most produce larvae; fecundity proportional to body size	Overgrowth; elevation of gape above substrate; gape movement prevents overgrowth; epizoism	High on shell	Shell, shell spines, ? toxins from dinoflagellates
Vermetidae	Encrusting; erect; or both when "solitary", erect and/or intertwined in massive "colonies"	(? apparently determinate but no reliable data)	Simple	Absent	Direct development or planktonic larvae	Overgrowth; erect growth habit to lift aperture from substrate; "colony" surfaces all apertures (not overgrown)	High on shells except for "colony" surfaces of shell apertures	Shell
Cirripedia	Encrusting (cemented); erect (stalked)	Determinate; exponentially decreases with age	Simple	Absent	Maturity at various ages (1 mo to >1 yr); produce larvae; occasional self-fertilization	Overgrowth; undercutting	High on shell; apertures often overgrown	Shell
Ascidiacea	Encrusting; erect	Determinate; exponentially decreases with age	Simple	Absent	Mature at various ages (3 wk to >1 yr); produce larvae	Overgrowth; elevation of siphons above substrate; ? toxins	High for species with hard tunic surface; low for species that slough off tunic surface (? these also with surface toxins)	Toxins
Colonial: Demospongiae	Encrusting; massive; erect	Determinate for massive and erect forms; indeterminate for encrusters with exponential increase in colony area	Edge extension by cell proliferation, immigration and displacement followed by division into functional units; not true budding	Buds and branches → new sponges; break up of encrusters into two or more sponges; gemmules → free larvae or attached sponges	Absent in many intertidal individuals; produce larvae; fecundity proportional to sponge volume and area occupied	Overgrowth; can survive overgrowth if any connection to surface; toxins; tough spicules; tough fibrous outer surfaces	Generally very low except for other sponges; fouled sponges always with outer skeletogenous surfaces, often reinforced with sand	Toxins; spicules; tough fibers

TABLE 1 (Continued)

Invertebrate	Morphology	Growth Pattern	Method of Growth	Asexual Reproduction	Sexual Reproduction	Competition Mechanisms	Susceptibility to Epizoan Recruitment	Defense against Predation
Hydroidea	Encrusting; erect; polymorphism of polyps	Usually indeterminate; exponential or linear growth of individual stolons; exponential growth of total stolon length and hydranths	Linear growth of stolons; budding of hydranths	Colony budding; regeneration from small pieces of stolon or coenosarc	Rapid maturity; fecundity proportional to colony area and/or no. hydranths; produce larvae or medusae	Toxins and/or growth inhibitors; overgrowth; tissue invasion; nematocysts; stolons survive overgrowth if hydranths exposed	High on thecae and stolons	Nematocysts and their toxins; ? tissue toxins
Scleractinia ...	Encrusting; massive; erect	Usually indeterminate; most rapid in small colonies; exponential increase in no. polyps and colony area; growth slower than most cnidarians due to calcification processes	Division (budding) of polyps	Colony budding of new polyps; increase in no. colonies by death of intermediate areas	Slow maturity (yr); produce larvae; fecundity may be proportional to no. polyps and colony area	Extracoelenteric feeding response; nematocysts and toxins; overgrowth	Only on exposed calcareous skeleton, therefore greater on undersurfaces of foliaceous forms than on massive colonies	Nematocysts and toxins; tentacles; skeleton (tissues can largely retract within skeleton)
Bryozoa	Encrusting; erect; massive; polymorphism of zooids	Usually indeterminate; exponential increase in colony area	Budding of zooids	Budding of zooids; brown bodies → regeneration of zooids	Very rapid maturity; no. ovisacs and fecundity proportional to colony area; produce larvae	Simple overgrowth; stolonal overgrowth (frontal and terminal kenozoids); frontal budding; spines and raised calcareous margins to prevent overgrowth; stolons can survive overgrowth if have connection to surface	High, especially for erect and/or spiny forms; support diverse epifauna	External skeletons; avicularia
Ascidiacea	Encrusting; massive; zooid polymorphism	Complex; usually determinate; or cyclical expansion and contraction of colony	Budding (often complex patterns of different zooid types)	Budding of zooids; colony fission; extrusion of buds → new colonies	May mature very rapidly (3 wk); may alternate with periods of asexual reproduction; no. mature zooids and fecundity proportional to colony area; produce larvae	Overgrowth; toxins	Low in most species	Toxins

(From Jackson 1977a. Copyright 1977 by the University of Chicago.)

habitat? How does species richness vary as environmental irregularity varies? Why are there more species at lower than at higher latitudes?

5. Stability. How does the system respond to external disturbances? What happens if new species invade?

6. Harvesting. What are the effects on these communities of the regular removal of biomass by organisms at higher trophic levels (grazing, predation, human manipulation)?

On a somewhat more abstract level:

7. Niche relations. To what extent are niche ideas appropriate for communities with strong spatial competition?

We will not comprehensively answer any of these questions here, but we will begin to answer most of them.

I.2. Spatial competition and the niche.

In the modern theory of competition for renewable resources, a central role is played by the concept of the niche; indeed, to the extent that competition theory has been a unifying force in ecology, this has been largely through the niche concept. A species' niche can be loosely defined as its place or role in the community in relation to other species. However, the reader is warned that the word is actually used in several different senses in the literature (see, for instance, Whittaker, Levin, and Root 1973).

An adequate introduction to the niche concept is beyond the scope of this volume, but excellent reviews of the subject are available elsewhere (Whittaker and Levin 1975; Pianka 1976).

Competition for space differs profoundly from competition for renewable resources: the dynamic of spatial competition, in which spatial mobility plays an indispensible role, is fundamentally different from the dynamic of resource competition and, as we shall see in the next section, the kinds of mathematical model used for resource competition do not make sense even as approximations when there is strong spatial competition; moreover, space is partitioned among competitors in a different way from the partitioning of resources. But niche ideas were originally developed in order to understand resource competition. One must, then, ask to what extent these ideas can be applied to spatial competition.

We review in this section existing ideas on defining niche variables for sessile communities.

First, it must be recognized that in nature space itself is seldom without structure: there are gradients of soil type, moisture, elevation, temperature, and so on. These gradients allow a variety of adaptive strategies, hence the coexistence of many different plant species (Whittaker 1967). However, such gradients are generally appreciable only over rather large distances, so that their effects are primarily on between-habitat diversity (Whittaker 1972). As the spatial area under consideration becomes smaller, these gradients become less pronounced and fewer species can coexist within that area.

Over a spatial area which encompasses only a single habitat type, most environmental gradients will have become negligible (indeed, this is essentially the definition of habitat). However, even within what is generally regarded as a single habitat there will often be subtle variations in soil type and elevation (microtopography), so that the space is still in effect heterogeneous, allowing the definition of some niche variables.

These small-scale variations have been extensively studied, for instance by Kershaw (1958); Anderson (1961a, 1961b); Grieg-Smith (1961); Kershaw (1962); Harper, Williams, and Sagar (1965); Yarronton and Green (1966); Zedler and Zedler (1969); Byer (1970); Brereton (1971); Hall (1971); Bratton (1976); and Harner and Harper (1976); and there can be no doubt that they often do play an important role in structuring plant communities.

Even an area of truly homogeneous space does not necessarily rule out the possibility of biotically defined niche variables. For instance, we pointed out in the preceding section that a plant needs a certain amount of space so that it can capture the sunlight in that space. But generally a plant will not be able to capture all the sunlight falling on the space that it occupies. For this reason we can regard the sunlight incident on each small area of the Earth's surface as a resource to be partitioned; plants can share this light by growing in several vertical strata, with plants in lower strata adapted to growth in less intense light (Lazenby 1955; Bjorkman and Holmgren 1966; Sparling 1967; Horn 1971; Whittaker 1972).

Such strata are characteristic of the foliage in forests (for instance, Kershaw 1973; Smith 1973) and are also seen in algal communities (Dayton 1975b; Menge 1976). Several authors (for instance, Whittaker 1972; Terborgh 1973) have suggested that a niche dimension could be based on such partitioning of light.

Root systems are often similarly stratified, and Parrish and Bazzaz (1976) have explicitly formulated a niche variable on this basis. For additional literature, see Mooney and Dunn (1970) and the works cited by Parrish and Bazzaz.

With regard to sessile fauna, there has been less emphasis on niche ideas; the possibilities do in fact seem more limited than for plants. However, Kohn and Leviten (1976) have shown that habitat structural diversity in tropical intertidal benches correlates significantly with species richness of predatory gastropod molluscs, suggesting that these sedentary creatures perceive their habitat as inhomogeneous, and Hart (1977) has shown that variations in the size and shape of rocks affect invertebrate benthic stream communities in much the same way that microtopographic variations affect plant communities.

Porter (1976) has suggested a biotically defined niche space for reef-building corals. These creatures combine autotrophic and heterotrophic functions, which yield the two axes of Porter's niche space: surface/volume ratio and polyp diameter, related to ability for light capture and zooplankton capture. Porter shows that Caribbean reef coral species can be ordinated in this space.

It is clear that niche ideas like these will play an important role in a final theory of spatial competition; in nature, and especially for plants, spatial competition and resource competition occur in subtle combination. However, our purpose here is to explore the possible importance of spatial competition in this combination. In order to do this we shall, as detailed in Section 4, restrict our attention to communities in which opportunities for niche differentiation are extremely limited. We shall take up the discussion of niche ideas again, from a more critical standpoint, in Section V.1.

I.3. Modelling competition for space.

An organism which needs space in which to live must be able to move about, in order to find and utilize any free space that may be available. (For instance, every sessile species has a dispersal phase in its life cycle: seeds, spores, and larvae are typical examples.)

Therefore, a theory of competition for space must include dispersal as one of its dynamical elements.

This can be achieved by means of a class of dynamical systems which we will call <u>reaction-dispersal</u> systems. (They are very often referred to as "reaction-diffusion systems" but, as pointed out by Levin (1978a), this terms should be reserved for a particular sort of dispersal.) These have been of use in a variety of contexts (for instance: Kolmogoroff, Petrovskij, and Piskunov 1937; Turing 1952; Othmer and Scriven 1971; Hlávacek, Sinkúle, and Kubícek 1972; Balsler and Degn 1975; Torre 1975; Nicolis and Prigogine 1977), and were brought into ecology by Skellam (1951) and Levin (1974).

Each population model applies to some <u>area</u> of space, in which the modelled populations live:

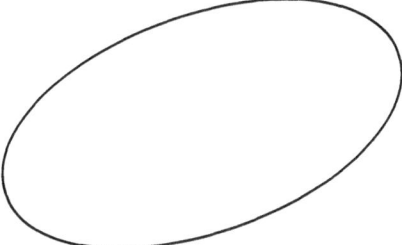

One generally deals with, as will we, an area which is small enough so that it is more or less uniform as a physical environment or habitat, but large enough so that the populations of interest are more or less contained within it (not too much emigration from or immigration to the area). These restrictions on the area can easily be relaxed with relatively minor changes to the mathematical formalism.

The set of all living things in such an area constitutes an ecological community. This same word "community" is also commonly used to refer to a subset of species which have in some way similar ecological requirements, and we will follow this practice here: for us, a community will consist of a set of mutual competitors for space.

Suppose the community of interest consists of n species. The dynamical variables usually adopted in population studies are the n number- or biomass-densities x_i, i=1,...,n: x_i is the number or biomass per unit area, i.e. the total number or biomass of species i divided by the total area. A typical population model is then a system of differential equations of the form

$$\text{(I.1)} \qquad \frac{dx_i}{dt} = f_i(x_1, \ldots, x_n) \ , \qquad i=1,\ldots,n \ .$$

But we can investigate the community more closely by dividing our area up into patches:

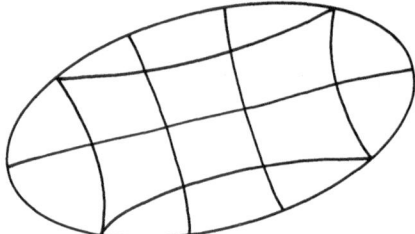

These patches may be naturally occurring units in the habitat (as, for instance, the rocks in a rocky subtidal zone, described by Osman 1977), or they may be artificial divisions imposed by the investigator (as in the quadrat samples of botanists, described for instance by Kershaw 1973).

In either case, we can now define for each species a density in each patch. If we do this, we will generally find that the densities in different patches are different; moreover, the spatial patterns of density thereby revealed will generally change as time progresses. These spatial patterns constitute additional community structure, not accessible to models of the form of equation (1).

The reaction-dispersal approach to this structure is as follows. Suppose there are m patches. We adopt as our dynamical variables the $m \times n$ quantities x_i^{μ}, $\mu=1,\ldots,m$, $i=1,\ldots,n$, where x_i^{μ} is the number- or biomass-density of species i in patch μ, that is the number or biomass of species i in patch μ divided by the area of patch μ. Then we write

$$\text{(I.2)} \qquad \frac{dx_i^{\mu}}{dt} = F_i^{\mu}(x_1^1, \ldots, x_n^m) = f_i^{\mu}(x_1^{\mu}, \ldots, x_n^{\mu}) + J_i^{\mu}(x_1^1, \ldots, x_n^m) \ ,$$

$$\mu=1,\ldots,m$$

$$i=1,\ldots,n.$$

The term f_i^μ is the growth rate of species i in patch μ in the absence of dispersal; these terms describe the effects of local interactions among species ("reaction"). The term J_i^μ is the net contribution to the growth rate of species i in patch μ due to dispersal of species i to and from patch μ.

The assumption that there is no migration to or from the total area covered by all the patches takes the form

$$(I.3) \qquad \sum_\mu J_i^\mu = 0 , \qquad i=1,\ldots,n.$$

We shall in this work assume that the within-patch dynamics is governed by Lotka-Volterra models of the form

$$(I.4) \qquad f_i^\mu = C_i x_i^\mu (k_i - \sum_j \alpha_{ij} x_j^\mu), \qquad \mu=1,\ldots,m, \qquad i=1,\ldots,n,$$

where C_i, k_i, and α_{ij} are positive constants. We choose without loss of generality $\alpha_{ii}=1$ for all i; then k_i is the carrying capacity of species i. The constant $C_i k_i$ is the intrinsic growth rate of species i, and the α_{ij} are the competition coefficients. For more detail about these models, see, for instance, the discussions of Slobodkin (1961), MacArthur (1972), and Maynard Smith (1974).

The vast majority of theoretical work on competition has used Lotka-Volterra models. A good justification for the use of such a model is to regard it as a low-order Taylor series expansion about some equilibrium point of a more general model. This makes sense if use is made only of properties of the model in a neighborhood of equilibrium. We will, however, make extensive use of global properties of the models and therefore cannot view them in this way. But these models do incorporate the essential features of competition, and the qualitative aspects of our results are obviously not strongly model-dependent. It is therefore reasonable to take these results as insights into the qualitative behavior of generic spatial competition models.

For the dispersal terms, we will use what Levin (1974) calls the "passive" form:

$$J_i^\mu = \sum_\nu D_i^{\mu\nu}(x_i^\nu - x_i^\mu)$$

(I.5)

$$= D_i \sum_\nu E_i^{\mu\nu} x_i^\nu , \qquad E_i^{\mu\nu} \geq 0 \; (\mu \neq \nu) , \qquad \sum_\nu E_i^{\mu\nu} = 0 ,$$

where $D_i^{\mu\nu}$, $E_i^{\mu\nu}$, and D_i are constants. The first form given here clearly shows the assumed dynamics of migration: it is assumed that each
species migrates independently of all the others, and that there is a
net flow of each species from patches with a relatively high density
of that species into patches with a relatively low density of that
species. Again, this is an idealization which is not expected to be
literally realistic (see, for instance, Levin 1976a, p. 10), but it
does embody what is usually the essential feature of dispersal: the
quest for free space.

In the second form given in equation (5), a more compact notation
is used for calculational convenience. Here the $E_i^{\mu\nu}$, which we shall
call the _dispersal_ _coefficients_, are taken to be of the order of unity,
and serve to specify the _range_ of dispersal for each species: $E_i^{\mu\nu}$ is
different from zero if and only if patches μ and ν are situated so that
a member of species i can move directly from patch ν to patch μ. The
quantities D_i, which we shall call _dispersal_ _parameters_, specify the
speed of dispersal for each species.

We can without loss of generality, and will throughout this study,
choose to express the species densities in units which have the property that all carrying capacities k_i are equal to 1. Thus, our models
have the form

(I.6) $$\frac{dx_i^\mu}{dt} = C_i x_i^\mu (1 - \sum_j \alpha_{ij} x_j^\mu) + D_i \sum_\nu E_i^{\mu\nu} x_i^\nu , \qquad \mu = 1, \ldots, m, \quad i = 1, \ldots, n,$$

until Chapter IV, where we will add a term to take harvesting into
account.

*

It is instructive to return to the usual global densities x_i and

to attempt to recover the usual models of the form (1). Assuming, for
the sake of simplicity, that all patches have the same area, the global
densities are

$$x_i = \frac{1}{m} \sum_\mu x_i^\mu \ .$$

Summing both sides of equation (6) over μ and using (3), we find

$$\frac{dx_i}{dt} = C_i x_i (1-\sum_j \alpha_{ij} x_j) - C_i \sum_j \alpha_{ij} \frac{m-1}{m} S_{ij} \ , \qquad i=1,\ldots,n \ ,$$

where

$$S_{ij} = \frac{1}{m-1} \sum_\mu (x_i^\mu - x_i)(x_j^\mu - x_j) \ , \qquad i,j=1,\ldots,n \ .$$

This is not of the form (1), but it can be approximated by an equation
of that form if the quantities S_{ij} are small enough to be neglected.

We can see when such an approximation makes sense if we regard
each set $\{x_i^\mu |$ i fixed, $\mu=1,\ldots,m\}$ as a statistical sample parametrized
by μ. Then S_{ii} is the variance of sample i and S_{ij}, $i \neq j$, is the co-
variance of samples i and j. (In models which are more general than
the Lotka-Volterra type, higher moments will also enter.) Thus S_{ii} is
a measure of the non-uniformity of the spatial distribution of species
i, and S_{ij}, $i \neq j$, is a measure of the influence of species i and j on
one another's spatial distribution. That is, large S_{ij}'s are an indi-
cation precisely of competition for space. In fact, these quantities
have been extensively used to quantify competition in the analysis of
botanical field data (Kershaw 1973).

We conclude that when there is strong competition for space the
usual models of the form (1), with global densities as dynamical variables, do not make sense even as approximations to the space-averaged
behavior of reaction-dispersal systems.

For further discussion of the formulation of these models, the
reader is referred to the works of Levin (1974, 1976a, 1976b, 1978a).

*

Probably the major shortcoming of these models is their neglect
of the age structure of the modelled populations. This is likely to
be especially important for sessile organisms, for dispersal, growth,
and competition tend to take place in different life phases of these
organisms (cryptophytes being an exception). However, there is in the
construction of ecological models an unavoidable tradeoff between generality and completeness: the more systems a model reasonably approxi-
mates, the less completely does it describe any one of them, and con-
versely.

The object of the present study is to arrive at a general picture
of the consequences of spatial competition for community structure,
and this entails the loss of much detail. Of course, studies like this
one should be complemented by more detailed studies of particular systems; for this purpose one could easily add age structure to the reaction dispersal framework. (For a somewhat different approach to this
structure, see the very interesting coral reef model of Maguire and
Porter 1977).

In those cases where the patches are artificially imposed divisions rather than naturally occurring units, some care is required in
the choice of patch size. It is well known, for example, that quadrat
size has major effects on the outcome of statistical botanical studies
(Pielou 1969; Goodall 1970; Kershaw 1973). For our purposes the smallest patch size within the bounds of practicality should in general be
a reasonable choice. This is supported, at least for plants, by studies on species-area curves, which find that these curves tend, for a
wide variety of plant communities, to level off at the smallest areas
censused (e.g., Hopkins 1955).

Indeed, one can in principle let the patch size shrink to zero
and use a continuous spatial description instead of the discrete one
adopted here. One would hope that the continuous and discrete versions
of the theory could be seen as equivalent descriptions of the same phenomena, but this may not be the case.

For instance, the discrete theory treated here has, for sufficiently small dispersal parameters, nontrivial stationary spatial patterns (Levin 1974; Chapters II and III below), but in similar (though
not entirely equivalent) continuous theories it appears that this may

not hold (see, for instance, Levin 1978a). Conway, Hoff, and Smoller (1977) and Conway and Smoller (1977) show that certain continuous theories do not admit nontrivial stationary patterns if the dispersal parameter is sufficiently large, which is, however, also true of at least some discrete theories (Levin 1974). This aspect of the theory needs clarification.

Since many of our examples are from plant communities, a word should be said about the measurement of population densities for plants. Traditionally botanists have tended not to think in terms of populations, largely because, as observed by J. L. Harper in his recent book (1977), it is difficult to decide what is a sensible measure of population density for plants. It is not always possible to distinguish plant "individuals", and even when this can be done there is typically so much variation in size among individuals that number density is seldom a useful quantity for plants.

An encouraging case can be made for using biomass density as a population variable (Harper 1977), but it has the disadvantage that it is usually very difficult to measure. Harper goes on to suggest that it may often be fruitful to measure the number density of _parts_ (for instance, leaves or shoots) of plants, and he gives some most interesting examples of this approach. We will assume here that there do exist appropriate densities, and that our equations are expressed in terms of these.

*

A related approach to spatial problems has been taken by Cohen (1970), Levins and Culver (1971), Horn and MacArthur (1972), and Slatkin (1974). The variables used in this approach are, in place of the population densities in each patch, simply the number of patches occupied by each species, or, more generally, the number of patches in each of a finite number of possible "states". This is a sort of slow time scale approximation: the system is viewed on a time scale in which local equilibrium is reached so rapidly that one can neglect the transient phenomena. Such models contain less information than the full reaction-dispersal models, but are more tractable. This approach should be particularly appropriate in treating long-range dispersal over a patchwork environment of similar habitats, rather than spatial structure within a single habitat area, as considered here.

I.4. Limitations of this study.

The study reported in this volume is carried out within the context of 1) very little possibility for niche differentiation, 2) slow
dispersal.

As discussed at length in Section 2, niche differentiation could
come about either through spatial homogeneity of the habitat or through
local coexistence. Both of these possibilities will be restricted
here, as follows.

1a) Spatial homogeneity. One could incorporate spatial heterogeneity into the models discussed in Section 3 by allowing the parameters
C_i and α_{ij} in our local Lotka-Volterra models to be functions of the
patch label μ. We have ruled out this possibility (and with it the
possibility of microtopographic niche differentiation) by taking these
parameters to be constants, independent of μ.

This also excludes from our discussion the class of phenomena
associated with alterations of the environment by organisms inhabiting
it. It is, for example, well known that plants affect the soil in
their neighborhood through depletion of nutrients, addition of organic
matter, and in a variety of other ways (cataloged nicely, with citations
from the literature, by Fox 1977, p. 83).

These alterations are responsible for some important phenomena,
namely obligatory successions (Section III.5) and the cyclic vegetational changes first noticed by Watt (1947) and described briefly in
Section III. 3. In order to bring these phenomena into the theory we
would have to allow the Lotka-Volterra parameters C_i and α_{ij} to depend
not only on the patch label μ, but on the past history of population
densities in each patch.

1b) Small probability of local coexistence. We shall restrict
(but not entirely rule out) the possibility of species coexistence
within a patch through local niche differentiation, by appropriate
choice of the competition coefficients α_{ij}.

One can interpret their numerical values by looking at the interaction between each pair of species in the community if all other species are absent and there is no dispersal. If, for example, we view
the interaction between two species i and j in this way, we find the
Lotka-Volterra system

$$\frac{dx_i}{dt} = C_i x_i (1 - x_i - \alpha_{ij} x_j)$$

$$\frac{dx_j}{dt} = C_j x_j (1 - x_j - \alpha_{ji} x_i).$$

It is well known (see, for instance, Slobodkin 1961, Emlen 1973, or Maynard Smith 1974) that such systems fall into three generic categories, depending upon the numerical values of the competition coefficients, as follows (recall that we have chosen $k_i = 1$ for all i):

$$\left.\begin{array}{l} \alpha_{ij} < 1 \\ \\ \alpha_{ij} < 1 \end{array}\right\} \Longrightarrow \quad \text{coexistence}$$

$$\left.\begin{array}{l} \alpha_{ij} > 1 \\ \\ \alpha_{ij} > 1 \end{array}\right\} \Longrightarrow \quad \text{contingent competition}$$

$$\left.\begin{array}{l} \alpha_{ij} < 1 \\ \\ \alpha_{ij} > 1 \end{array}\right\} \Longrightarrow \quad \text{species i dominates species j}$$

In the case of coexistence, there is a stable equilibrium with both species' densities positive. If there is contingent competition, then the only equilibrium with both densities positive is a saddle point, so the outcome of the competition is contingent upon the initial values of the densities. If species i dominates species j, then species j will be eliminated no matter what the initial values of the densities (unless the initial density of species i is zero). For further details, the books of Slobodkin (1961), Emlen (1973), and Maynard Smith (1974) can be consulted.

We will eventually construct communities by choosing the competition coefficients α_{ij}, $i \neq j$, independently and at random from some

positive real interval (e,f), with a probability distribution which is
uniform over this interval. Let

$$
p = \begin{cases} (1-e)/(f-e) \ , & e<1 \\ \\ 0 & , \ e \geqslant 1 \end{cases}
$$

be the probability for a competition coefficient, so chosen, to be less
than 1. Then the probability that any one of these randomly chosen
two-species interactions is coexistent is given by p^2. We shall always
take the interval (e,f) so that p is less than or equal to 0.1. Then
the probability for a randomly chosen pair of species to coexist locally (by means of some sort of niche differentiation) is never greater
than 0.01.

One could of course imagine that the probability distribution of
the alphas is non-uniform over the interval (e,f), but our results are
obviously highly robust to the form of this distribution, and a uniform
distribution is the simplest possibility. One could also impose some
sort of functional dependence among the alphas. For example, there is
some justification for assuming that the alphas are symmetric, $\alpha_{ij} = \alpha_{ji}$,
when they express pure resource competition (MacArthur 1972). But competition for space involves a strong component of interference (recall
Section 1), in which case there is no reason whatever for the alphas
to be symmetric.

2) _Slow dispersal_. We shall assume in addition that the interpatch dispersal of all species in the community is "slow", in the sense
that the dispersal parameters D_i are "small" for all i. In fact, for
the sake of simplicity we shall take

$$D_i = D = \text{constant} > 0 \text{ for all i}$$

with D "small". We shall henceforth refer to D as _the dispersal para-
meter_.

This assumption restricts our treatment to sessile and sedentary
competitors for space. Mathematically, it justifies the application

of perturbation methods (Appendix A) to these systems; thus, strictly speaking, when we say that D is "small" we mean it is sufficiently small so that our various perturbation expansions in powers of D converge.

By taking all the dispersal parameters to have the same numerical value D we have lost part of the distinction between fugitive and equilibrium species, but for our purposes this distinction will not play an essential role. Recall (Hutchinson 1951) that (to put it shortly though somewhat crudely) fugitive species are poor competitors but rapid dispersers, while equilibrium species are good competitors but slow dispersers. In the situations that we will consider here, any fugitive species would be eliminated from the system at an early stage in its development; in particular, in all our calculations in which the magnitude of D is essential only equilibrium species will remain in the system.

We can, then, save ourselves some work without losing any essential content by taking all the dispersal parameters equal. This procedure would no longer make sense if we were to treat patchy disturbance such as damage from wind, fire, or waves — then the relative rates of dispersal would matter very much and fugitive species would be an important element of the community — but we shall neglect such influences here, except for a brief discussion in Section IV.5.

3) Form of the dispersal coefficients. Recall that the dispersal coefficients $E_i^{\mu\nu}$ serve to specify the range of dispersal of species i. It is extremely difficult to write down realistic dispersal coefficients, and we will not attempt this here. Rather, we will consider two extreme possibilities for these coefficients, hoping thus to get an idea of the range of behavior to be expected. In addition, we shall choose the dispersal coefficients to be the same for all species:

$$E_i^{\mu\nu} = E^{\mu\nu}$$

for all i.

In the extreme of maximum mobility, all patches are equally accessible. This is expressed in terms of the quantities $D_i^{\mu\nu}$ (which occur in the first form of equation (5)) as follows:

$$D_i^{\mu\nu} = \begin{cases} D, & \nu \neq \mu \\ 0, & \nu = \mu. \end{cases}$$

The corresponding values for the dispersal coefficients are

$$(I.7) \qquad E^{\mu\nu} = \begin{cases} 1, & \nu \neq \mu \\ 1-m, & \nu = \mu, \end{cases}$$

where m is the number of patches. We shall call this <u>random dispersal</u>.

In the extreme of minimum mobility, we imagine that all the patches are lined up in a (one-dimensional) row, and that each species is constrained to move linearly through the row, with no possibility of jumping over patches. Then

$$D_i^{\mu\nu} = \begin{cases} D, & \nu = \mu-1 \\ D, & \nu = \mu+1 \\ 0, & \text{otherwise,} \end{cases}$$

so that

$$(I.8) \qquad E^{\mu\nu} = \begin{cases} 1, & \nu = \mu-1 \\ -2, & \nu = \mu \\ 1, & \nu = \mu+1 \\ 0, & \text{otherwise} \end{cases}$$

We shall call this <u>linear nearest neighbor dispersal</u>.

By taking the dispersal coefficients to be the same for all species
we are losing some of the rich biological detail that characterizes
many of these communities; here again we are sacrificing detail for
generality. We will, however, discuss briefly one effect of differing
ranges of dispersal in Section III.6.

I.5. Colonization.

We assumed in Section 3 that the populations of interest to us are
more or less contained within the area of space to which our model applies; that there is not too much excahange of biomass between this
area and .other habitat areas outside it. Of course, it would not make
sense to assume a rigorous lack of all such exchange: our assumption
is that there is "not too much" exchange in the sense that the net flow
of biomass of any species into any patch of our area from outside the
area is negligible compared with the net flow from other patches of the
area. This is a reasonable assumption for, say, a fairly isolated forest or mussel bed, under normal circumstances.

There is, however, a special circumstance to which this assumption
does not apply. Suppose that our area has been in some way cleared of
life, so that it consists simply of empty space. Then whatever immigration there is from outside the area constitutes, at first, the entire population dynamics. Our area will be colonized. Eventually the
area will be fully occupied and then our assumption that interpatch
dispersal is far more important than immigration makes sense.

We discuss here neither the origins of these colonizers nor their
mechanisms for long-range dispersal (which are sometimes different from
their mechanisms for dispersal within a community), since an excellent
account of these matters, with citations from the literature, can be
found in Chapter 14 of Emlen's book (1973).

We shall, in Chapters II, III, and IV, attempt to characterize the
communities which arise, under various conditions, from such colonization of empty space. In mathematical terms, the process of colonization will serve to specify the initial conditions to be imposed on the
system (6), as follows.

We assume that at first our area is unoccupied. Recall that it has
been partitioned into m patches. We assume next that n species arrive
from outside the area, colonizing it in such a way that each patch is

monopolized by one species, which attains a density in that patch near its carrying capacity. We call the corresponding values of the densities x_i^μ colonization initial values, and let the system evolve from these values in accordance with equations (6).

We can describe this mathematically by giving a function c: $\{1,\ldots,m\} \rightarrow \{1,\ldots,n\}$, which we call the colonization function, such that for each $\mu \in \{1,\ldots,m\}$, $c(\mu) \in \{1,\ldots,n\}$ is the species which initially colonizes patch μ. Then the corresponding colonization initial values are

(I.9) $x_i^\mu = \delta_{ic(\mu)}$, $i=1,\ldots,n,$ $\mu=1,\ldots,m,$

where it is to be recalled that we have normalized all carrying capacities to unity. Here δ_{ij} is the Kronecker delta

(I.10) $\delta_{ij} = \begin{cases} 1, & i=j \\ 0, & i \neq j . \end{cases}$

*

This is, to be sure, a simplified picture of colonization: in reality there will usually be a period of time during which the colonization and internal dynamical processes are mixed together. However, it is intuitively obvious that, for the situations considered in this volume, the final outcome will be essentially the same as that obtained from our simple idealization. One would have to be much more cautious in treating patchy disturbances, as from fire, wind, and waves: then the relative rates of disturbance and recolonization would be important. However, we shall, except for a brief discussion in Section IV.5, neglect such influences here.

*

Colonization is, of course, very strongly influenced by random effects, so that the particular colonization function c for each particular system is obtained, mathematically, as a sample function for some random function. We will consider two colonization processes in this volume.

One possibility, which we call _uniform colonization_, is that the colonizing species divide the patches among themselves more or less equally, with any two species occupying about the same number of patches, and with the patches occupied by each species randomly distributed in space. That is, the probability distribution for each patch is uniform over all the colonizing species, and these distributions are independent.

This case corresponds to simultaneous arrival of all the colonizers; in addition to idealizing a situation which frequently occurs in nature, it is a convenient conceptual and mathematical point of reference.

Our other form of colonization is closely related to the "niche pre-emption" hypothesis of Motomura (1932) and Whittaker (1965, 1972), and we shall call it colonization by _space pre-emption_. In this case, different species arrive at different times, with the first species to arrive appropriating a fraction k of the available patches (chosen at random), the next species to arrive appropriating a fraction k of the _remaining_ patches, and so on until all the patches are occupied.

This yields the geometric form of abundance-rank curve (Section III.1) seen in many terrestrial plant communities (Whittaker 1965), especially at early successional stages (Whittaker 1972; Bazzaz 1975).

I.6. Preview.

We sketch here the content of the remainder of this volume.

When one considers the development of these systems from colonization initial values, it quickly becomes evident that they divide naturally into three broad categories, depending upon the values of the competition coefficients α_{ij} ($i \neq j$).

If all these coefficients are greater than 1, then all pairwise interactions among the colonizers are contingent (Section 4); there are no coexistent or dominant interactions. This implies, for small D, that once a species has become established in a patch it can maintain predominance there: other species will be present in this patch in small numbers, but there will be no large-scale invasions from other patches. The structure of the community is determined primarily by the original colonization episode, and we call such a community <u>founder controlled</u>. These communities are discussed in Chapter II.

If some competition coefficients are greater than 1 and some less, then all three kinds of pairwise interactions enter, though coexistent interactions are rather unusual because of our assumption (Section 4) that opportunities for local niche differentiation are limited. In this case, a colonizer cannot necessarily maintain predominance in a patch; after colonization there will be a "reshuffle" of species among the patches, with some species being eliminated from the community entirely. This process is governed primarily by dominance relations among the species, and we refer to the resultant communities as <u>dominance controlled</u>. Most sessile communities seen in nature are of this type. These communities have a rich structure, which we explore in Chapter III.

There is a third type of community, in which all competition coefficients are smaller than 1. Coexistence in such a community will presumably depend very strongly on niche relations, so these communities fall outside the domain covered in this volume. Some work has been done on this case by Casten and Case (1978); see Section V.1.

In Chapters II and III only competitive relations are taken into account. But real space-limited communities are also subject to grazing or predation pressure from higher trophic levels. The effects of such pressure, which we subsume under the general heading <u>harvesting</u> along with intentional or unintentional removal of biomass by human action, on community structure are discussed in Chapter IV. Particular attention is devoted to the effects of harvesting on species richness, for there is an extensive natural historical literature on such effects. This literature is reviewed and compared with theory in Section IV.4.

In Chapter V the general conception of spatial competition which emerges from this study is first very briefly summarized, and then some implications of this way of thinking for several broad problems of theoretical ecology are explored.

Most of the more mathematical aspects of the discussion are relegated to appendices, particularly Appendix A, which also contains the derivation of formulae for the lowest order perturbations to semisimple eigenvalues of arbitrary complex matrices which are analytic in the perturbation parameter. This derivation is quite straightforward, but, so far as this author is aware, it has not appeared elsewhere in print.

Chapter II. FOUNDER CONTROLLED COMMUNITIES

We assume in this chapter that every pair of colonizing species
has (in the terminology of Section I.4) a contingent interaction, so
that $\alpha_{ij} > 1$ for all i, j (i≠j) with our convention $k_i = 1$ for all i. We
call a community formed from such colonizers <u>founder</u> <u>controlled</u>, because its structure is governed almost entirely by the colonization
process rather than by dynamical processes intrinsic to the community.

Most sessile communities documented in the natural historical
literature appear to be dominance controlled (Chapter III) rather than
founder controlled. However, as we shall see in Chapter IV, heavy
harvesting can sometimes induce founder control. Moreover, as detailed
in Section V.3, there are both theoretical and observational indications of founder control in some species-rich wet lowland tropical
forests.

II.1. Colonization equilibria.

We consider an area of originally empty space which is colonized by n species as described in Section I.5, and we assume that the
subsequent development of the community is governed by a system of differential equations of the form (I.6), with all $\alpha_{ij} > 1$ (i≠j). The initial values of the population densities x_i^μ are, then, given by colonization initial values

(II.1) $x_i^\mu = \hat{x}_i^\mu = \delta_{ic(\mu)}$, $\mu = 1, \ldots, m$, $i = 1, \ldots, n$,

where c is the colonization function of Section I.5.

This is not an equilibrium of the system (I.6). Therefore, as the effects of interpatch dispersal become felt the population densities will change. However, the values (1) of the densities do define an equilibrium of the system obtained from (I.6) by setting D=0 in that equation. Since we are assuming D to be small, we expect that there is an equilibrium of the full system (I.6) not very far from the initial values (1). If there is such an equilibrium, and if all population densities are nonegative at this equilibrium and it is stable, we expect that the population densities will approach the equilibrium asymptotically.

This is a problem in perturbation theory. Requisite concepts from this theory are sketched in Appendix A. (For our present purposes we will need mainly the material in Section A.1.)

Applying Proposition A.1 to the unperturbed equilibrium (1) gives sufficient conditions for the existence of an equilibrium which continuously approaches the initial values (1) as D approaches zero. If we denote by $F_i^\mu(x_1^1,\ldots,x_n^m;D)$ the right hand side of equation (I.6), then the first condition is that these functions should be analytic in all m×n+1 variables. This condition is stronger than we really need at this point, but it is satisfied and we will need it later on.

The second condition is that the matrix

$$(II.2) \qquad A^{(0)\mu\nu}_{ij} = \frac{\partial F_i^\mu}{\partial x_j^\nu}(\hat{x};0)$$

should be nonsingular. The eigenvalues of this matrix are calculated in Section A.3.1, and they are all nonzero; therefore $A^{(0)}$ is nonsingular.

Hence there is an equilibrium of the system (I.6) which continuously approaches the values (1) as D approaches zero. We will call it a __colonization equilibrium__.

We can calculate the values of the densities at such a colonization equilibrium to first order in D by using equation (A.5). We find in this way for the equilibrium densities

$$(II.3) \qquad x_i^\mu(D) = \hat{x}_i^\mu - \sum_{\nu,j} (A^{(0)-1})^{\mu\nu}_{ij} F_j^{(1)\nu}(\hat{x})D + \cdots$$

where the matrix $A^{(0)-1}$ is given by equation (A.36) and

$$F^{(1)\,\nu}_{\ \ j}(\hat{x}) = \sum_\sigma E^{\nu\sigma}_j \delta_{jc(\sigma)}.$$

Evaluating this expression, using equations (B.4, 5), yields finally

$$
(\text{II.4}) \quad x^\mu_i(D) =
\begin{cases}
- \displaystyle\sum_{\nu:c(\nu)=i} T_i(c(\mu))E^{\mu\nu}_i D \ , & i \neq c(\mu) \\[2ex]
1 - \left[\displaystyle\sum_{\nu:c(\nu)=c(\mu)} U_{c(\nu)}(c(\mu))E^{\mu\nu}_{c(\nu)} + \right. \\[2ex]
\qquad \left. + \displaystyle\sum_{\nu:c(\nu)=c(\mu)} (T_{c(\mu)}(c(\mu))+U_{c(\mu)}(c(\mu)))E^{\mu\nu}_i\right]D & , i = c(\mu)
\end{cases}
$$

where

$$
(\text{II.5}) \quad
\begin{cases}
T_i(c(\mu)) = -1/C_i(\alpha_{ic(\mu)}-1) \ , & i \neq c(\mu) \\[2ex]
U_i(c(\mu)) = \alpha_{c(\mu)i}/C_i(\alpha_{ic(\mu)}-1) \ , & i \neq c(\mu) \\[2ex]
T_{c(\mu)}(c(\mu))+U_{c(\mu)}(c(\mu)) = -1/C_{c(\mu)} .
\end{cases}
$$

In order for all the equilibrium densities to be positive, at
least to first order in D, we need $T_i(j)<0$ for all i, j (i$\neq$j), which
is equivalent to $\alpha_{ij}>1$ for all i, j (i$\neq$j). But this condition on the
competition coefficients was precisely our starting point in this chapter.

Therefore all population densities are positive to first order in
D at a colonization equilibrium. (For another way to prove this, see
Levin 1976a, Appendix.) This means that the densities are positive
for sufficiently small values of D: if D becomes too large, the first-
order approximation (4) no longer makes sense. However, we will find
in Section 2 that the colonization equilibria become unstable when D
is still quite small and the perturbations to the densities are quite
small. Therefore, so long as the colonization equilibria are stable

the approximation (4) makes sense and the densities are positive.

It is extremely plausible to assume that if the colonization equilibrium corresponding to our initial values (1) is stable then the system will approach this colonization equilibrium, and we adopt this assumption.

*

The basic structure of a founder controlled community is, then, for sufficiently small D, as follows.

In the absence of external disturbances such a community will generally be found near an equilibrium state, which will be a configuration such that in each patch one species is far more abundant than any other (this is made quantitative in the following section); it is simply the species which first colonized that patch. Thus, the relative abundances of the various species in the community will depend almost entirely upon accidents of colonization: there will not be "dominant" species typifying the communities found in similar habitats (as for example in the "Beech-Maple Association" so often seen in the American midwest (Williams 1936)). In general one would expect that most species will have comparable abundances.

The spatial distribution of each species among the patches will be quite random, so that any "clumping" or spatial pattern (Kershaw 1973) will be on a very small scale (of the order of the size of individual organisms). Species-area curves (Pielou 1969, 1975; May 1975) will increase monotonically with area even at very large areas.

The vast majority of sessile communities documented in the natural historical literature do not fit this description; instead, one almost always sees something very much like the dominance controlled communities that we will discuss in Chapter III. There is, however, one notable exception: the species-rich wet lowland tropical forest. Such a forest is characterized by precisely the attributes we have just listed. We will discuss the relevant observations in Section V.3, where we will also indicate why these forests are indeed more likely candidates for founder control than are temperate zone forests.

In the remainder of this chapter we examine the mathematical structure of founder control more closely. (Biologically-oriented readers may want to proceed now to Chapter III.)

II.2. Stability.

Local stability of an equilibrium can conveniently be characterized by the quantity

(II.6) $\Lambda = -\text{maximum}\{\text{Re}(\lambda_i^\mu)\}$,

where λ_i^μ ($\mu=1,\ldots,m$, $i=1,\ldots,n$) are the eigenvalues of the linearized
stability matrix (2) of the equilibrium. (For any complex number z,
Re(z) denotes the real part of z.)

In case $\Lambda>0$, the equilibrium is asymptotically stable in the sense
of Liapunov (LaSalle and Lefschetz 1961; Hahn 1967; this is often called "neighborhood stability" in the ecological literature). Moreover,
in this case the magnitude of Λ contains additional information, for
Λ^{-1} is of the order of the time scale for exponential return of the
system to equilibrium if it has been moved away from equilibrium by
some small disturbance.

Alternatively, if random environmental fluctuations are represen-
ted by taking some parameter in the model to be a random variable, Λ
can be used to formulate a sufficient condition for the system to remain within the domain of attraction of the equilibrium in question,
despite these random environmental fluctuations (May 1973; Yodzis 1978).

In Section A.3.1, Λ is calculated to first order in D (it is for
this calculation that we need the analyticity requirement of Proposition A.1) and its mean value over certain statistical ensembles of randomly assembled systems is evaluated numerically by Monte Carlo techniques.

We refer the interested reader to Section A.3.1 for details. In
this section we will discuss the significance of these calculations for
the magnitudes of the equilibrium populations and for the species richness of these communities.

Consider, for example, a community of 20 species. We can use
figures A.1-3 to find the upper bound D_{max} on D such that the average
colonization equilibrium is stable, and then use equations (4,5) to
calculate the range (x_{min}, x_{max}) of possible equilibrium values for the
population densities in a patch, for species other than the one that

Table II.1. Random dispersal.

(e,f)	D_{max}	(x_{min}, x_{max})
(1.1,1.9)	0.001	(0.005,0.100)
(1.5,2.5)	0.002	(0.007,0.042)
(3.0,4.0)	0.002	(0.002,0.007)
(10.0,20.0)	0.002	(0.0005,0.0022)

Table II.2. Linear nearest neighbor dispersal

(e,f)	D_{max}	(x_{min}, x_{max})
(1.1,1.9)	0.013	(0.008,0.138)
(1.5,2.5)	0.062	(0.021,0.124)
(3.0,4.0)	0.102	(0.017,0.051)
(10.0,20.0)	0.154	(0.004,0.017)

first colonized the patch. The results are given in tables 1 and 2, for several choices for the interval (e,f) from which the competition coefficients are randomly chosen to assemble a community (Section I.4, item 1b)).

We see that the values of D which permit stability and of the per- turbations to the equilibrium populations in this stable regime are both quite small. This justifies the use of first-order perturbation calculations for these communities in the stable regime. In particu- lar, it justifies our use of these methods to prove feasibility of the equilibria (that is, positivity of all equilibrium densities) in the preceding section.

We close this section by briefly inquiring to what extent the species richness of these communities is influenced by the local sta- bility of colonization equilibria. In this regard, one must distin- guish three domains of values of the dispersal parameter D, as follows. For D sufficiently small, say in an interval $(0,D_1)$, the colonization equilibria are stable for all n. As D increases into an interval (D_1,D_2), stability depends on n, and for D larger than D_2 colonization equilibria are unstable for all n>1.

Now, the domain (D_1,D_2) in which local stability depends on species richness turns out to be so small, and consequently any such dependence is so sensitive to the value of D, that this dependence is most unlikely to have any significance: it will in a real system be masked by random effects not explicitly included in these models. We can estimate the interval (D_1, D_2) by using the average values of Λ calculated in Section A.3.1. For example, if $(e, f) = (3.0, 4.0)$ and dispersal is random, we obtain $(D_1, D_2) = (.0013, 0.0029)$ from figures A.1 and A.2.

So the species richness of these communities is governed by extrinsic factors such as colonization and disturbance when dispersal is slow enough so that our perturbation methods apply.

II.3. Total standing crop.

The total number of individuals or of biomass (depending on how we have chosen to define our densities) in one of these communities is

$$x_T = A \sum_{\mu,i} x_i^\mu \; ,$$

where we have assumed, for the sake of simplicity, that all patches have the same area A. To save writing, we choose units of area in which $A = 1$. We will call x_T the total standing crop.

If the system is at a colonization equilibrium with colonization function c, we find from equations (4,5)

$$x_T = m + \left[\sum_{\substack{\mu,\nu \\ c(\mu)=c(\nu)}} C_{c(\mu)}^{-1} E_{c(\mu)}^{\mu\nu} - \sum_{\substack{\mu,\nu \\ c(\mu)=c(\nu)}} \frac{{}^\alpha c(\mu) c(\nu)^{-1}}{C_{c(\nu)} ({}^\alpha c(\nu) c(\mu)^{-1})} \times \right. $$
$$ \left. \times E_{c(\nu)}^{\mu\nu} \right] D$$

to first order in D.

We consider communities which are assembled at random from a pool of species available for colonization, and express this by choosing each parameter α_{ij} at random from a positive real interval (e, f) with uniform probability distribution over this interval, and each parameter

C_i at random from a positive real interval (g,h) with uniform probability distribution over this interval, as explained in Section I.4,
item 1b).

Consider the set of all communities assembled in this way, for
some fixed pool of colonizers, hence fixed intervals (e,f) and (g,h),
and fixed numbers n of colonizing species and m of patches. The mean
value of x_T over this set of communities is

$$\langle x_T \rangle = m + \frac{\log(h/g)}{h-g} \left[\left\langle \sum_{\substack{\mu,\nu \\ c(\mu)=c(\nu)}} E^{\mu\nu}_{c(\mu)} \right\rangle \right. -$$

$$\left. - ((e+f)/2-1) \frac{\log((f-1)/(e-1))}{f-e} \left\langle \sum_{\substack{\mu,\nu \\ c(\mu)=c(\nu)}} E^{\mu\nu}_{c(\nu)} \right\rangle \right] \quad D$$

to first order in D.

Suppose the dispersal is random, so that the dispersal coefficients E have the form (I.7). For any particular colonization function c, let m_i, $i \in \{1,\ldots,n\}$, be the number of patches μ such that
$c(\mu)=i$. Then of course

$$\sum_i m_i = m.$$

Now

$$\sum_{\substack{\mu,\nu \\ c(\mu)=c(\nu)}} E^{\mu\nu} = \sum_{\mu} \sum_{\nu:c(\nu)=c(\mu)} E^{\mu\nu} = \sum_{\mu}((m_{c(\mu)}-1)+(1-m)) =$$

$$= \sum_{\mu}(m_{c(\mu)}-m) = m_{c(1)}+m_{c(2)}+ \cdot \cdot \cdot +m_{c(m)}-m^2 = \sum_i m_i^2-m^2$$

and

$$\sum_{\substack{\mu,\nu \\ c(\mu)=c(\nu)}} E^{\mu\nu} = \sum_{\mu} \sum_{\nu: c(\mu)=c(\nu)} E^{\mu\nu} = \sum_{\mu}(m-m_{c(\mu)}) = m^2 - \sum_i m_i^2.$$

Suppose first that the colonization is uniform (Section I.5). Then in each patch the probability of colonization by species i is $1/n$, independent of the patch and of i, where n is the number of colonizing species. Thus the probability distribution of the m_i is a multinomial distribution (for instance, Johnson and Kotz 1969) with all probabilities $p_i = 1/n$. Therefore (Johnson and Kotz 1969), for any i

$$\left\langle m_i^2 \right\rangle = \frac{m(m-1)}{n^2} + \frac{m}{n}$$

so

$$\left\langle \sum_i m_i^2 \right\rangle = \frac{m(m-1)}{n} + m$$

and

$$\left\langle \sum_{\substack{\mu,\nu \\ c(\mu)=c(\nu)}} E^{\mu\nu} \right\rangle = -\left\langle \sum_{\substack{\mu,\nu \\ c(\mu)=c(\nu)}} E^{\mu\nu} \right\rangle = m(m-1)\left(1-\frac{1}{n}\right).$$

Finally, then, for random dispersal and uniform colonization,

(II.7) $$\left\langle x_T \right\rangle = m - \frac{\log(h/g)}{h-g}\left[((e+f)/2-1)\frac{\log((f-1)/(e-1))}{f-e}+1\right]m(m-1)\left(1-\frac{1}{n}\right)D$$

If the colonization is by space pre-emption (Section I.5), then the probability p_i for colonization by species i is (e.g., Whittaker 1965)

$$p_i = (1-k)^{i-1}k,$$

where k is the fraction of still vacant patches colonized by each species as it arrives. Therefore

$$\left\langle \sum_i m_i^2 \right\rangle = m(m-1)\frac{k}{2-k} + m,$$

so that for random dispersal and colonization by space pre-emption one has again equation (7), but with the single modification

(II.8) $(1-\frac{1}{n}) \longrightarrow (1-\frac{k}{2-k})$.

We can estimate the expected number of species in the community for a given k by taking the largest value of i such that mp_i is greater than 1. We find in this way that the factor $1-k/(2-k)$ is a monotonically increasing function of n.

The results should be substantially the same for linear nearest neighbor dispersal, with the coefficient of D in equation (7) smaller by a factor of the order of $1/m$. On the other hand, the upper bound on D which permits stability is in this case at least an order of magnitude greater (table 2).

                                    *

We conclude that for sufficiently small dispersal parameters, 1) average total standing crop decreases as the dispersal parameter increases, 2) average total standing crop decreases as the number of species in the community increases.

The author is unaware of any data from founder controlled communities upon which these results impinge. However, they should also hold to a reasonably good approximation for the dominance controlled communities discussed in the next chapter, and we will compare them with data from such communities in Section III.4.5.

II.4. Summary.

This chapter discusses communities which are formed from colonizing species for which $\alpha_{ij} > 1$ for all i, j (i≠j), under our convention

k_i =1 for all i.

In Section 1 it is shown that after colonization such a community tends, if the dispersal parameter is sufficiently small, toward a "colonization equilibrium", which is such that in each patch one species is far more abundant than any other; it is simply the species which first colonized that patch. Thus the relative abundances of the various species in the community depend almost entirely on accidents of colonization, rather than on factors intrinsic to the community dynamics. Other characteristics of these communities are briefly discussed and it is pointed out that many species-rich wet lowland tropical forests display just these characteristics.

Section 2 summarizes some calculations, carried out in detail in Appendix A, of local stability for colonization equilibria. These results are used to justify the use of perturbation calculations for the population densities when the colonization equilibria are stable. It is also argued that local stability does not have a significant influence on species richness.

Finally, in Section 3, the total standing crop of biomass in these communities is averaged over certain ensembles of randomly assembled communities. It is found that average total standing crop is a monotonically decreasing function of the dispersal parameter D and of species richness. These calculations will be used in the following chapter to estimate the corresponding quantities for dominance controlled communities, and will be compared there with observations.

Chapter III. DOMINANCE CONTROLLED COMMUNITIES

We relax here the restriction of Chapter II to purely contingent pairwise interactions between species: in this chapter we treat communities which are assembled from species for which the competition coefficients α_{ij} ($i \neq j$) are randomly chosen from a real interval (e,f) with $0 < e < 1 < f$. This allows the possibility also of dominant or of coexistent pairwise interactions. However, as explained in Section I.4, we restrict the probability of a coexistent interaction between two randomly chosen species to be always less than 0.01, so that there is little possibility for local niche differentiation.

We call the resulting communities <u>dominance controlled</u>, because their structure is governed primarily by interspecies dominance relations. These communities have a rich structure which, as we shall see, mirrors much of the structure to be seen in space-limited communities in nature.

III.1. Numerical dominance and functional dominance.

The word "dominance" is commonly used in three distinct ways in ecology. One of these, social dominance, which is an intraspecific phenomenon epitomized by the "pecking order" among hens (Schjelderup-Ebbe 1935), will not concern us here. Our interest will be engaged rather by the two forms of interspecific dominance.

One of these we will call <u>numerical dominance</u>. If one measures (in terms, for instance, of number, biomass, coverage, or some combination of these) the abundance —— sometimes called "importance" in this context —— of each species in a community, one will sometimes find that one or a few species are far more "important" than all others: these species are <u>numerical dominants</u> for that community.

Strong numerical dominance is characteristic of temperate zone terrestrial plant communities, a fact around which the early "plant sociologists" such as du Rietz (1930) and Braun-Blanquet (1932) built up a considerable body of work (see also Whittaker 1962). Communities of sessile fauna also tend to exhibit strong numerical dominance (see, for instance, Paine 1974).

The most thorough expression of these numerical relations is the abundance-rank diagram for a community: this is a graph of the abundance of each species on the vertical axis (most often with a logarithmic scale) versus the rank of that species in the sequence of species from most to least abundant, on the horizontal axis (see, for instance, Whittaker 1965, 1970; and figures 6, 7, and 8 below). Such a graph is thorough but cumbersome; for this reason ecologists have long sought a single numerical measure of the degree of numerical dominance in a community.

In fact, several such "dominance indices" have been suggested. They (and the related "evenness" or "equitability" indices) are a part of the vast literature concerned with numerical measures for the "diversity" of a community. To discuss these indices here would carry us too far afield, and in any case admirable discussions are already to be found elsewhere (for instance, May 1975 and Pielou 1975).

It will be convenient for our purposes to have the following special notion of numerical dominance. If the densities x_i^μ in some patch μ are all either small (meaning of the order of the dispersal parameter D) or large (meaning of the order of the carrying capacity), then the species with large densities will be called predominant species in patch μ. This terminology will be used frequently in what follows.

The other form of interspecific dominance, which we will call functional dominance, is not so directly empirical as numerical dominance, but is nevertheless extremely useful. If two species S_1 and S_2 are competitors, then S_1 functionally dominates (or sometimes in the literature: competitively dominates) S_2 if the two cannot coexist in a small space, and no matter what the (nonzero) initial densities of the two species, it is always S_1 that drives S_2 to local extinction.

A paradigm for functional dominance is provided by a Lotka-Volterra model

$$\frac{dN_1}{dt} = C_1 N_1 (k_1 - N_1 - \alpha_{12} N_2)$$

$$\frac{dN_2}{dt} = C_2 N_2 (k_2 - \alpha_{21} N_1 - N_2)$$

with $\alpha_{12} < k_1/k_2$ and $\alpha_{21} > k_2/k_1$. The zero isoclines for such a system are shown in figure 1. The reader can easily see that every nontrivial

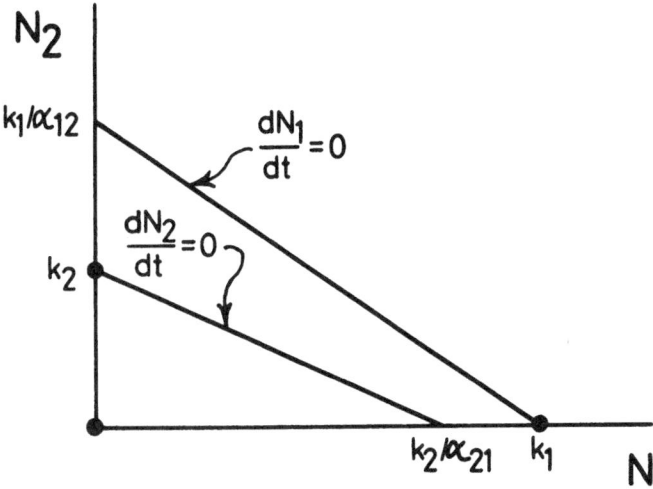

Figure III.1. A Lotka-Volterra model for functional dominance of species 2 by species 1.

trajectory tends to the equilibrium at $N_1 = k_1$, $N_2 = 0$. (For a more axiomatic and general discussion, see Yodzis 1976a.)

Although Gause's (1934) first experiments in competition demonstrated functional dominance of one yeast species over another, functional dominance has generally played a less important role than other forms of competitive interaction in mathematical ecology. However, in spatial competition ―― and therefore in this book ―― functional dominance is of central importance.

This functional dominance is achieved in a variety of ways. Among primary producers, morphology can give an overwhelming advantage in competition for light (Horn 1971; Dayton 1975a, 1975b). Root morphology in terrestrial plants is, as noted in Section I.1, of great competitive significance, and it can confer functional dominance (McCowan and Williams 1968). Allelochemicals often win dominance for the species that produce them, among both plants (Keever 1950; Muller 1966; Overland 1966; Rice 1968; Gant and Clebsh 1975) and sessile animals (Jackson and Buss 1975).

Osmotic interference and myrmecophy, both discussed in Section I.1, are less common mechanisms of dominance. Some corals achieve dominance by overtopping and digesting their competitors (Lang 1973; Porter 1974), and the physical aggression commonly seen in intertidal communities often confers dominance (Connell 1961a; Dayton 1971, 1975b; Paine 1974). Rapid growth and a flexible attachment mechanism account for the functional dominance of some mytilids (Paine 1971).

Our analysis will show, unsurprisingly, that numerical dominance and functional dominance often go hand in hand. But this need not always be the case: a functional dominant may be kept down to rather small numbers by predation (Pearson 1942; Dayton 1972; Section IV.3.2), and of course numerical dominance in a founder controlled community has nothing to do with functional dominance.

III.2. The reshuffle.

In Chapter II we studied the communities which result when free space is colonized by species having only contingent pairwise interactions (Section I.4) with one another. Our object in the present chapter is to see what sorts of communities result when the colonizing species are allowed the possibility of dominant pairwise interactions (and, with much smaller probability, coexistent interactions) as well. We shall, however, now restrict our attention only to the case of random dispersal (equation (I.7)) and of uniform colonization (Section I.5), except for a few remarks in Section 6 about other forms of dispersal.

Consider, then, a colonization episode as described in Section I.5, which produces the colonization initial values

$$x_i^\mu = \hat{x}_i^\mu = \delta_{ic(\mu)} \quad , \quad \mu=1,\ldots,m, \quad i=1,\ldots,n,$$

with colonization function c, from which the community thus originated
evolves in accordance with a system of equations of the form (I.6):

$$(III.1) \quad \frac{dx_i^{\mu}}{dt} = C_i x_i^{\mu}(1-\sum_j \alpha_{ij} x_j^{\mu})+D\Sigma E_i^{\mu\nu} x_i^{\nu} \quad , \quad \mu=1,\ldots,m, \quad i=1,\ldots,n$$

with the dispersal coefficients E given by equation (I.7).

The corresponding colonization equilibrium has the densities
(II.4,5). But now we are assuming that some of the alphas are less
than 1; therefore some of the densities at the colonization equilibrium
are negative. It would, then, be absurd to suppose now, as we did in
Chapter II, that the system will approach this equilibrium: the densities can never evolve to negative values. It is to be expected rather
that we will now have a "reshuffle" of species among the patches, with
some species being eliminated from the community entirely.

Consider first a single patch μ, populated initially by species
c(μ). At first the growth rate in this patch of another species i≠c(μ)
will be, from equation (1),

$$\frac{dx_i^{\mu}}{dt} = D \sum_{\nu:c(\nu)=i} x_i^{\nu}.$$

That is, the densities of all other species will initially grow. While
these densities are still small, we have to first order in small quantities (small meaning here, as always in this study, of the order of
D),

$$(III.2) \quad \frac{dx_i^{\mu}}{dt} \approx C_i x_i^{\mu}(1-\alpha_{ic(\mu)}) + D \sum_{\nu:c(\nu)=i} x_i^{\nu}.$$

There are now two possibilities:

1) In case $\alpha_{ic(\mu)}>1$, the first square bracket in equation (2) is
negative and will eventually cancel out the second (positive) square
bracket. Therefore the density of species i will remain small in patch

μ. This will be the case either if species i and species c(μ) have a contingent pairwise interaction, or if species c(μ) dominates species i. (Recall Section I.4, item 1b), and our condition on $q_{ic(\mu)}$.)

2) In case $\alpha_{ic(\mu)}<1$, the first squate bracket in (2) will be, like the second square bracket, positive. This means that the growth of the density of species i in patch μ will not necessarily stop at a small value. Rather, species i may be able to _invade_ the patch: that is, to become predominant (Section 1) in it. This will be the case either if species i and species c(μ) have a coexistent interaction, or if species i dominates species c(μ).

Let

(III.3) $I(c(\mu)) = \{i \mid \alpha_{ic(\mu)}<1\}$

be the set of all _potential_ _invaders_ of a patch in which c(μ) is predominant. These potential invaders will be inhibited by species c(μ) and by one another from invading; in general they will not all be able to invade the patch.

One cannot say with certainty what will happen without actually solving the system (1) of differential equations, which is out of the question. However, we can formulate a plausible hypothesis as to what will happen, and look at the implications of this hypothesis for the qualitative behavior of various properties of the resultant communities, averaged over an appropriate ensemble of communities as in Section II.3. In order for this to give us a reasonable picture of these communities, it suffices if our hypothesis holds (merely) most of the time.

Now, the growth of a potential invader i is inhibited by the term

$$-\sum_{j}\alpha_{ij}x_{j}^{\mu}$$

in equation (1). The larger the α_{ij}, the greater will be this inhibitory effect. For each of our systems we will choose the α_{ij} (i≠j) randomly from an interval (e,f) of real numbers, with 0<e<1<f in order to allow the possibility of non-contingent interactions. Then the proba-

bility that an α_{ij} $(i \neq j)$ is less than 1 is given by

(III.4) $p = (1-e)/(f-e)$.

If this probability is small enough, then it is reasonable to suppose that the probability for more than one potential invader to succeed is negligible. We shall take the probability p always less than or equal to 0.1, and we will assume that this is "small enough" in the sense just described. (For instance, we will see later, in figure 2, that the probability for a stable equilibrium with three species present is then always less than 0.001.)

We have defined p as the probability that a randomly chosen competition coefficient is less than 1. We see from (3) that it is also the probability that a randomly chosen species can potentially invade a patch which is predominated by some other randomly chosen species. This is its more fundamental interpretation; moreover, this interpretation is not tied to any particular choice (for instance, Lotka-Volterra) of local dynamics. The probability p will play a central role in the remainder of this volume.

From equation (2) we see that the potential invader for which the quantity $1-\alpha_{ic(\mu)}$ is the largest will grow the most quickly at small densities. It seems reasonable to assume that this potential invader has the best chance of invading. Let us, then, adopt

Hypothesis III.1a. If the set I(s) of potential invaders of some patch in which species s is predominant is nonempty, then that potential invader i for which α_{is} is the smallest, and only that potential invader, will successfully invade the patch.

Such an invasion can take two forms:

1) If $\alpha_{si} > 1$, then species i dominates species s (Section I.4). We see from equation (1) that, if $c(\mu) = s$, then when x_i^μ has attained values of the order of the carrying capacity 1, the growth rate of species $s = c(\mu)$ will in patch μ be negative. The density of species s in patch μ will decline to some small value; species s will be replaced by species i as the predominant species in patch μ.

2) If $\alpha_{si} < 1$, then species i and species s can coexist as the (two) predominant species in the patch.

*

However, the process does not necessarily end there. We must next consider whether there are potential invaders of a patch predominated by (in case 1) species i or (in case 2) species $c(\mu)$ and i together. If there is only one predominant species we can proceed again as we just did; if there are two predominant species, essentially the same considerations apply, but with the following alterations.

The growth rate at low densities of a potential invader i of a patch predominated by two species s and t is governed by the quantity

$$(III.5) \quad 1-\alpha_{is}\hat{x}_s(t)-\alpha_{it}\hat{x}_t(s),$$

where $\hat{x}_s(t)$ is, for two species s and t with a coexistent interaction, the value of x_s in the Lotka-Volterra two-species equilibrium:

$$(III.6) \quad \hat{x}_s(t) = \frac{1-\alpha_{st}}{1-\alpha_{st}\alpha_{ts}} .$$

Therefore, the set of potential invaders is now

$$I(s,t) = \{i \mid \alpha_{is}\hat{x}_s(t)+\alpha_{it}\hat{x}_t(s)<1\}$$

and our hypothesis takes the form

Hypothesis III.1b. If the set $I(s,t)$ of potential invaders of some patch in which species s and t are predominant is nonempty, then that potential invader i for which $\alpha_{is}\hat{x}_s(t)+\alpha_{it}\hat{x}_t(s)$ is the smallest, and only that potential invader, will successfully invade the patch.

In general, these invasions may of course result in patches with three predominant species. The likelihood of this is estimated in figure 2, which depicts the outcome of a Monte Carlo simulation in which three-species Lotka-Volterra systems were constructed at random, with the competition coefficients α_{ij} (i≠j) chosen randomly with a probability p that $\alpha_{ij}<1$ (i≠j). For each p the fraction of systems having a stable equilibrium with all three population densities positive was

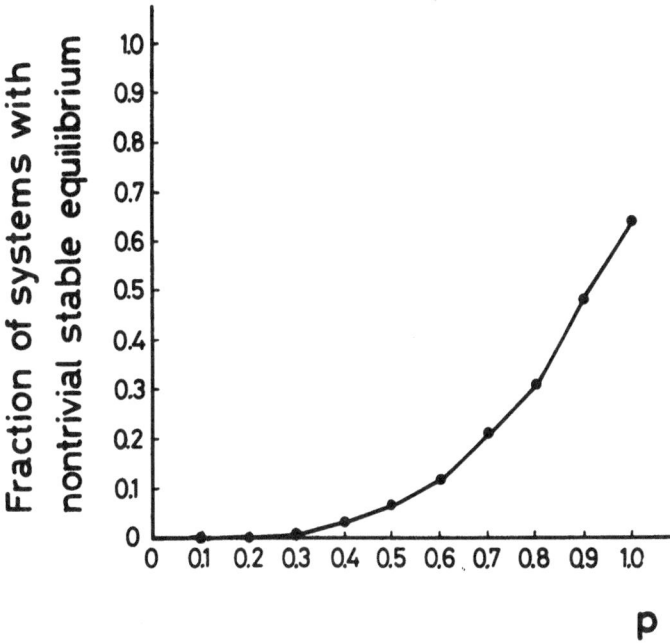

Figure III.2. The fraction of three-species
Lotka-Volterra systems with a nontrivial stable
equilibrium as a function of the probability p
that a randomly chosen competition coefficient
is less than 1.

counted. For each p, 1000 systems were examined.

It is seen from figure 2 that with the restriction $p \leqslant 0.1$ that we
have already adopted earlier in this section, the probability of a
stable equilibrium with all three population densities positive is very
small (in fact it is less than 0.001). Therefore it is reasonable to
assume, when $p \leqslant 0.1$, that if a three-species Lotka-Volterra system has
a stable equilibrium with fewer than three positive population densities, the system will approach this equilibrium.

Notice also that if $i \in I(s,t)$, then either $\alpha_{is} < 1$ or $\alpha_{it} < 1$. This
follows directly from the circumstance that for every stable two-spe-
cies Lotka-Volterra equilibrium, $\hat{x}_s(t) + \hat{x}_t(s) > 1$.

Therefore, an invasion by species j of a patch predominated by
species s and t will take one of the following forms:

i) If $\alpha_{js}>1$, $\alpha_{jt}<1$, and

a) $\alpha_{tj}<1$ and $\alpha_{sj}\hat{x}_j(t)+\alpha_{st}\hat{x}_t(j)>1$, then the only stable equilibrium with fewer than three positive densities is that with species j and t coexisting. So in this case species j and t will predominate.

b) $\alpha_{tj}>1$ and $\alpha_{sj}>1$, then the only stable equilibrium with fewer than three positive densities is that with only species j having a non-zero density. So in this case species j will dominate both species s and species t, and species j will predominate.

ii) If $\alpha_{js}<1$, $\alpha_{jt}>1$, and

a) $\alpha_{sj}<1$ and $\alpha_{tj}\hat{x}_j(s)+\alpha_{ts}\hat{x}_s(j)>1$, then species j and s will predominate.

b) $\alpha_{sj}>1$ and $\alpha_{tj}>1$, then species j will predominate.

iii) If $\alpha_{js}<1$, $\alpha_{jt}<1$, and

a) $\alpha_{sj}<1$, $\alpha_{tj}>1$, and $\alpha_{tj}\hat{x}_j(s)+\alpha_{ts}\hat{x}_s(j)>1$, then species j and s will predominate.

b) $\alpha_{sj}>1$, $\alpha_{tj}<1$, and $\alpha_{sj}\hat{x}_j(t)+\alpha_{st}\hat{x}_t(j)>1$, then species j and t will predominate.

c) $\alpha_{sj}<1$, $\alpha_{tj}<1$, and

A) $\alpha_{tj}\hat{x}_j(s)+\alpha_{ts}\hat{x}_s(j)>1$ and $\alpha_{sj}\hat{x}_j(t)+\alpha_{st}\hat{x}_t(j)<1$, then species j and s will predominate.

B) $\alpha_{tj}\hat{x}_j(s)+\alpha_{ts}\hat{x}_s(j)<1$ and $\alpha_{sj}\hat{x}_j(t)+\alpha_{st}\hat{x}_t(j)>1$, then species j and t will predominate.

C) $\alpha_{tj}\hat{x}_j(s)+\alpha_{ts}\hat{x}_s(j)>1$ and $\alpha_{sj}\hat{x}_j(t)+\alpha_{st}\hat{x}_t(j)>1$, then species j and either species s or species t, depending upon the initial conditions, will predominate. When this case occurs in our calculations, we choose species s or species t at random (with equal probability) to predominate along with species j.

d) $\alpha_{sj}>1$ and $\alpha_{tj}>1$, then species j will predominate.

iv) Otherwise, there is no stable equilibrium with fewer than three positive densities. With our restriction $p \leqslant 0.1$ this case will arise only very seldom. When it does arise in our calculations, we will throw the system in which it occurs out of our statistical sample.

*

Continuing in this way, we obtain a sequence of single-species invasions into each patch; the set of all these sequences for all patches in our area is what we call a <u>reshuffle</u>.

There are two possibilities: either the sequence of invasions terminates in every patch after a finite number of steps, with no more invasions possible, or not. We shall investigate each of these two possibilities in the following two sections.

III.3 Quasicycles and lotteries.

Suppose there is at least one patch in which the sequence of invasions just described never terminates. Then the system never settles down into an equilibrium; in at least one patch there will be an infinite sequence of invasions.

An infinite sequence of invasions must be cyclic, for there are only a finite number of combinations of species which can predominate in a patch (with our restriction to no more than two predominant species the number of possible combinations is $n(n+1)/2$, where n is the number of colonizing species), so that after a finite number of steps the predominant species will be the same as in some earlier step in the sequence, and then the sequence will have to repeat itself. But then after a finite number of further steps it will have to repeat itself again, and so on.

So the time development of the system will be something like cyclic. We shall call such a time development a <u>quasicycle</u> (rather than a cycle), because it need not be asymptotic to a limit cycle, as the following considerations show.

To each sequence of invasions into a patch there corresponds a sequence of predominant species in that patch. Consider two such sequences of predominant species, from two different patches. Obviously, if these two sequences have any one element in common, then the two sequences are identical from that element on. If this is the case, we will say that these two sequences of predominant species are <u>equivalent</u>; otherwise we will call them <u>inequivalent</u>.

If for some system all the infinite sequences are equivalent then it is reasonable to suppose, as have Gilpin (1975) and May and Leonard

(1975) in discussing a very similar behavior for a somewhat simpler system, that the system is in its time development approaching a limit cycle.

But not every quasicycle can be asymptotic to a limit cycle. Let C be the largest number of inequivalent infinite sequences of predominant species among the set of all such sequences for some system, and suppose that C>1. Then it is still reasonable to suppose that the projection of the system development into the subspace of the system phase space spanned by the patches linked by any maximal set of equivalent sequences is cyclic. But even so, if the periods of these projected developments are not rational multiples of one another the time development of the system as a whole will not be cyclic; instead of approaching a limit cycle the time development will fill out an entire C-dimensional torus contained in the system phase space. The asymptotic set is then a "strange attractor" (Ruelle and Takens 1971), quite similar to the nonlinear double pendulum example discussed by Arnold and Avez (1970) (see also May and Oster 1976).

Quasicycles can occur if, for example, the dominance relations in a community are nontransitive. In the simplest example there would be three species A, B, C such that A dominates B, which dominates C, which dominates A. Jackson and Buss (1975) have found such competitive relationships among the epifauna which inhabit the undersurfaces of foliaceous corals such as Agaricia and Montastrea spp. in Caribbean reefs. However, quasicycles themselves have not yet been observed in these habitats, or elsewhere.

Is this an oddity, or can one expect quasicycles to occur frequently in nature? We can estimate the likelihood for quasicycles to occur as follows.

First, for each of several choices for a real interval (e,f), 100 systems of the form (1) are constructed by choosing the competition coefficients α_{ij} (i≠j) at random from the interval (e,f), and by choosing the Lotka-Volterra parameters C_i at random from the interval (1,2) (see Section I.4, item 1b)), using a random number generator. The number n of colonizing species is also varied, but the number m of patches is kept fixed at 200. The patches are colonized uniformly (Section I.5), again using a random number generator.

For each of these systems and for each initial colonization, the reshuffle process is simulated on a computer. Those few systems in

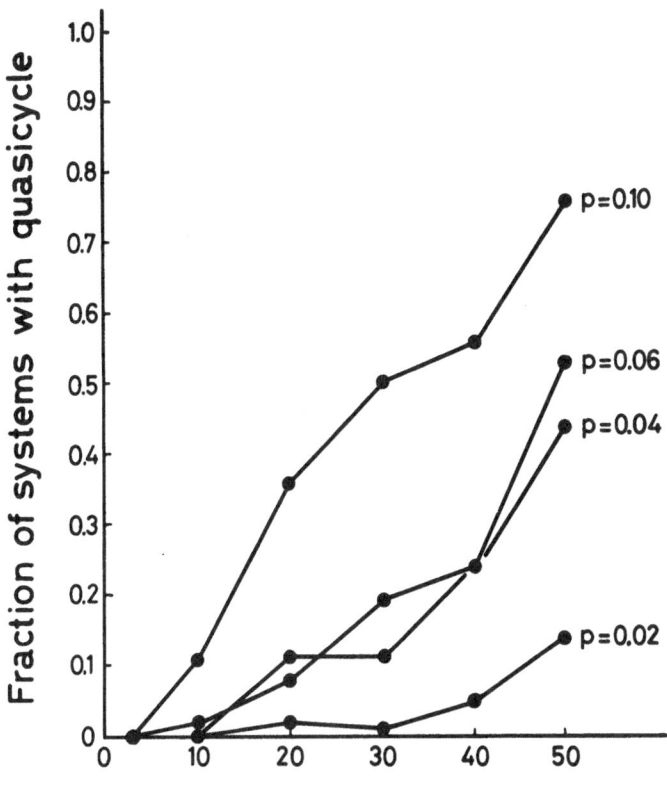

Figure III.3. The fraction of randomly assembled
systems having a quasicycle as a function of the
number n of colonizing species, for several values
of the probability p (equation (4)), obtained by
Monte Carlo simulations, as described in the text.

which a patch with three predominant species appears in the course of
the reshuffle are not used in the calculations. If for some system
the reshuffle has in some patch not terminated within n(n+1)/2 invasions (the number of different combinations of n species chosen 1 or
2 at a time), that system has a quasicycle. For each (e,f) and for
each n the number (of the original 100) systems having a quasicycle is
counted.

The results are depicted in figure 3. These results suggest that quasicycles may be quite likely for communities with a rich supply of colonizing species. However, this probability is overestimated in figure 3, for several reasons.

1. Figure 3 is based on the assumption of random dispersal, that is, very high mobility of all species. In nature, not all species are so mobile —— but in order to keep a quasicycle going, all species in the network must have a certain minimum mobility.

2. R. T. Paine points out (private communication) that the area under consdieration must be large enough to accomodate all the species involved in a network in order for quasicycles to be realized in that area. This is especially relevant to systems such as that studied by Jackson and Buss (1975), which are isolated, relatively small habitat islands (see above, and Section 4.2).

3. Our Hypothesis 1 (Section 2) may, for several reasons, fail, and one may have instead of a quasicycle a "lottery". This will be discussed at the end of this section.

Despite these reservations, the suggestion remains here that the behavior of species-rich systems may often be qualitatively different from that of species-poor systems. Another suggestion along these lines comes from Holling (1973). He points out that species-rich systems can be expected to have many different stable equilibria, each with its domain of attraction, while species-poor systems are more likely to have relatively few domains of attraction. (Computer simulations of randomly constructed Lotka-Volterra systems with different numbers of species done by Gilpin and Case (1976) lend support to this idea, as does the obvious possibility for the systems studied here to have many different colonization functions, many of them leading to different stable equilibria.)

Holling goes on to reason that, while a species-poor system with only one or two domains of attraction is likely to experience an extinction if it is disturbed in some way, a species-rich system, with its many domains of attraction, has a good chance of merely being moved into a different domain of attraction by a random disturbance. Thus, although each equilibrium of a species-rich system will tend to be less stable than a typical equilibrium of a species-poor system (as shown by May 1974), the species-rich system will in its global behavior be (in Holling's terminology) more "resilient" —— that is, roughly, more constant in species composition. In addition to the examples cited by

Holling (1973), the work of Sutherland (1974) on marine fouling communities seems to illustrate these ideas in the field.

*

The cyclic vegetational changes observed by Watt (1947, 1955),
Nagel (1950), Schaeffer and Moreau (1958), and Lieth (1960) are not
quasicycles, though their mechanism is quite similar to that of quasicycles. In these "Watt cycles", each species alters the environment
in its immediate neighborhood so as to favor invasion by the next species in the cycle, these invasions taking a cyclic form exactly like
our cyclic interpatch invasions. As already pointed out in Section
I.4, the form we have chosen for our local "within-patch" dynamics does
not encompass such environmental alterations by the resident biota.
For a thorough discussion of Watt cycles, see the recent article by Fox
(1977).

*

The existence of quasicycles depends upon Hypothesis 1 of Section
2, or at least on some sort of predictability as to which species will
invade which patch. There are two ways in which this could break down.

Firstly, it may be that the question of which potential invader is
successful depends not only on the predominant species in a patch (as
we have assumed), but also on the particular (small) densities of all
the other species in the patch. This seems rather unlikely so long as
our assumption $p \leqslant 0.1$ holds, but if this assumption were dropped it
could very well occur. Such a dependence could be quite sensitive to
these densities; then successful invasion would be highly sensitive to
random environmental influences, so that the identity of the successful
invader would be largely a matter of chance.

Secondly, the dispersal process itself may be subject to such random influences, so that the fate of each patch depends not so much on
the dynamics within that patch as on which potential invader happens to
arrive there first. Again, the identity of the successful invader
would be largely a matter of chance.

In either case, we would have instead of a deterministic sequence
of invasions into each patch a random sequence of invasions. We will

call this kind of behavior a <u>lottery</u>. Observations by Sale (1975) of
the spatio-temporal behavior of a guild of territorial coral reef fish
suggest something very much like a lottery. In a later paper, Sale
(1978) suggests the term "lottery" for this behavior, and offers a
somewhat different explanation for it.

*

To summarize: detailed spatio-temporal observations of communities
where there is strong competition for space should eventually reveal
quasicycles in some species-rich, relatively undisturbed communities.
However, random effects will often produce lotteries instead.

III.4. Equilibrium communities.

We consider in this section those systems for which, in the reshuffle of Section 2, the sequence of invasions in every patch terminates after a finite number of steps, with no more invasions possible.

In such a case the number of species in the community will at
first steadily increase as more and more species arrive, and then
steadily decrease to some equilibrium value as more and more species
are eliminated in the reshuffle. This behavior is seen in many studies
of succession (for references, see Section 5), and in a particularly
beautiful portrayal Dayton (1973b) has actually recorded such a process
in detail, showing the arrival into and departure from a rocky intertidal community of each species. This can be seen in figure 4, which is
adapted from a similar figure of Dayton (1973b).

In each of several replicates, an area of substratum was scraped
clean, and the additions to (denoted with a plus sign in front of the
abbreviated species name) and subtractions from (denoted with a minus
sign) the community were recorded at one-month intervals. Only the
most abundant species are treated: "presence" of a species means
presence in more than half of the replicate plots. In the experiment
which lead to figure 4, predators were excluded by means of wire cages
in order that one could observe the effects of pure competition for
space.

We see that in this case even our idealization (Section I.5) of a
colonization episode followed by a reshuffle quite faithfully mirrors

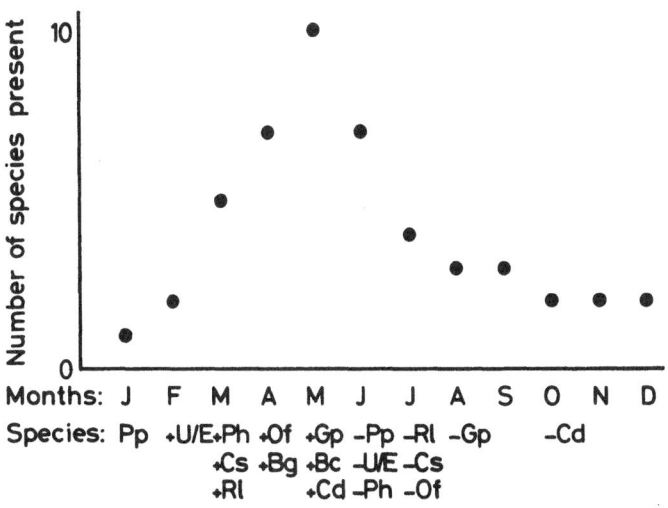

Figure III.4.   Species which were present in over
half of the treated plots in the intertidal zone
of San Juan Island, Washington (USA).   The abbre-
viations used for species names are as follows:
Pp=Porphyra perforata, U/E=Ulva or Entermorpha
(not differentiated in the field), Bg=Balanus
glandula, Cd=Chthamalus dalli, Gp=Gigartina pa-
pillata, Mc=Mytilus californianus, Hg=Hallosac-
cion glandiforme, Ha=Heterochordaria abietina,
Cs=Colpomenia sinuosa, Em=Endocladia muricata,
Bc=Balanus cariosus, Ph=Polysiphonia hendryi,
Rl=Rhodomela larix, Of=Odonthalia floccosa.
(After Dayton 1973b.   Copyright 1973 by The University of Chicago.)

reality: for the first five months only additions to the community are
seen, and after that only subtractions from it. However, it can be expected that sometimes the colonization and reshuffle processes will be
mixed temporally.

*

At the end of the reshuffle each patch is predominated by one or
two species; if there are two predominant species in some patch, then
the two have a coexistent interaction with one another. Denote the
predominant species in a patch μ by $s(\mu)$ and $t(\mu)$, with the convention

(III.7) $t(\mu)=0$ if there is only one predominant species in patch μ.

It will be convenient for us to define the quantities

$$\alpha_{0i}=\alpha_{i0}=0 \ , \quad i=1,\ldots,n$$

(III.8)

$$\alpha_{00}=0 \ , \quad k_0=0 .$$

Because any pair of species which predominate together in the same patch must have a coexistent interaction with one another, at the end of the reshuffle the values of the densities are near the values

(III.9) $x_i^{\mu}=\hat{x}_i^{\mu}=\delta_{is(\mu)}\hat{x}_{s(\mu)}(t(\mu))+\delta_{it(\mu)}\hat{x}_{t(\mu)}(s(\mu))$,

where $\hat{x}_s(t)$ is, for any s and t, given by equation (6). Because of the definitions (7), (8) this reduces to the usual single-species equilibrium in each patch with only one predominant species.

The values (9) for the densities define an unperturbed equilibrium of our system (that is, an equilibrium of (1) with D=0; cf. Section II.1). Reasoning as in Section II.1, we can show that there is a corresponding perturbed equilibrium —— an equilibrium of the full system (1) with D nonzero but small —— near the unperturbed equilibrium (9). It is extremely reasonable to suppose that if this perturbed equilibrium has all species densities nonnegative and if it is stable, then the system will asymptotically approach this equilibrium. We adopt this supposition here, and proceed to scrutinize its consequences. As in Chapter II, we will use the perturbation methods discussed in Appendix A.

III.4.1. Equilibrium populations.

The perturbed equilibrium corresponding to (9) is again given by equation (II.3),

(III.10) $x_i^\mu(D) = \hat{x}_i^\mu - \sum_{\nu,j} (A^{(0)-1})_{ij}^{\mu\nu} F^{(1)\nu}_j(\hat{x})D,$

where now

$$(A^{(0)-1})_{ij}^{\mu\nu} = \delta_{\mu\nu} A_{ij}^{-1}(s(\mu),t(\mu)),$$

and, since $\hat{x}$ is now given by (9),

$$F^{(1)\nu}_j(\hat{x}) = \sum_\xi E_j^{\nu\xi} \hat{x}_j^\xi = \sum_\xi E_j^{\nu\xi} (\delta_{js(\xi)} \hat{x}_{s(\xi)}(t(\xi)) + \delta_{jt(\xi)} \hat{x}_{t(\xi)}(s(\xi))).$$

The Lotka-Volterra matrix $A^{-1}(s(\mu),t(\mu))$ is given by equations
(B.4) and (B.5) if $t(\mu)=0$ (that is, if there is only one predominant
species in patch μ), and by (B.15), (B.16), (B.17), (B.18) if $t(\mu) \neq 0$.
However, under the conventions (7) and (8) all of the two-species formulae of Appendix B automatically reduce to the corresponding one-
species formulae when there is only one predominant species. For instance, if $t(\mu)=0$, then (B.15)-(B.18) reduce to (B.4) and (B.5).
Therefore, we can simply use the two-species formulae throughout the
following, with no need for special mention of the patches with one
predominant species.

In order to check that none of the equilibrium populations are
negative, we must substitute the appropriate quantities from Appendix
B into equation (10). The resultant explicit expressions for $x_{s(\mu)}^\mu$
and $x_{t(\mu)}^\mu$ are complicated and not very illuminating; it is in any case
clear that these densities are positive for sufficiently small D. For
$i \neq s(\mu), t(\mu)$ we find

$$x_i^\mu(D) = C_i^{-1} (\alpha_{is(\mu)} \hat{x}_{s(\mu)}(t(\mu)) + \alpha_{it(\mu)} \hat{x}_{t(\mu)}(s(\mu)) - 1)^{-1} \times$$

$$\times \sum_{\xi:s(\xi)=i} E_i^{\mu\xi} \hat{x}_{s(\xi)}(t(\xi)) + \sum_{\xi:t(\xi)=i} E_i^{\mu\xi} \hat{x}_{t(\xi)}(s(\xi))\ D.$$

The expression in square brackets is manifestly positive. To investigate positivity of the expression in front of the square brackets, we

recall that the reshuffle has at this point terminated, and therefore $\alpha_{is(\mu)}>1$ and $\alpha_{it(\mu)}>1$ for all μ and for all species i (other than $s(\mu)$ and $t(\mu)$) which remain in the system, not having been eliminated in the reshuffle. (If these conditions are not met, then further invasions are possible (Section 2), and the reshuffle has not terminated.) So

$$\alpha_{is(\mu)}\hat{x}_{s(\mu)}(t(\mu))+\alpha_{it(\mu)}\hat{x}_{t(\mu)}(s(\mu)) > \hat{x}_{s(\mu)}(t(\mu))+\hat{x}_{t(\mu)}(s(\mu))$$

$$=\frac{2-\alpha_{s(\mu)t(\mu)}-\alpha_{t(\mu)s(\mu)}}{1-\alpha_{s(\mu)t(\mu)}\alpha_{t(\mu)s(\mu)}} \quad .$$

Because all pairs $s(\mu)$, $t(\mu)$ with $t(\mu)\neq0$ have coexistent interactions and because of (8), $\alpha_{s(\mu)t(\mu)}<1$ and $\alpha_{t(\mu)s(\mu)}<1$ for all μ. It follows immediately that the last expression is greater than 1.

Therefore, to first order in D all species densities are nonnegative at the perturbed equilibrium. We will check the reliability of this first-order approximation in Section 4.4.

III.4.2. Species richness.

In the course of a reshuffle some species will be eliminated from the community entirely: they will be functionally dominated by some other species, but will themselves be unable to invade another patch.

In the computer simulations of the reshuffle process described in Section 3, for those systems in which the reshuffle terminated after a finite number of steps the number of species remaining in each system after the reshuffle was counted. These numbers were then averaged for each choice of the number n of colonizing species and of the probability p for $\alpha_{ij}<1$ ($i\neq j$) (equation (4)).

The results of these computations can be seen in figure 5. The horizontal axis is the number n of colonizing species and the vertical axis is the average number of species remaining after the reshuffle, for those reshuffles that terminated after a finite number of steps. This relationship is plotted for several different values of p. Error bars on the curve for p=0.020 are for one standard deviation.

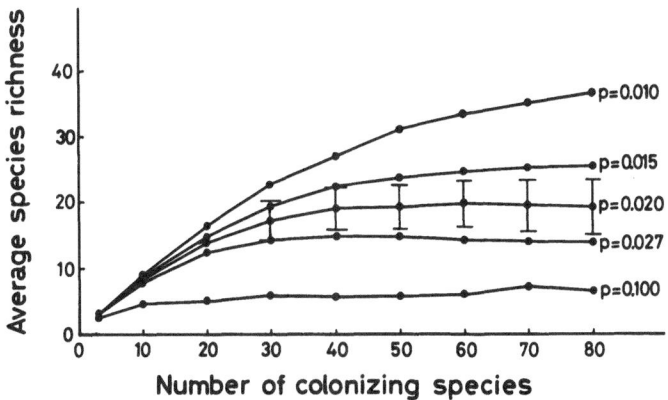

Figure III.5. The average species richness of
reshuffled equilibria as a function of the number of colonizing species, for several values
of the probability p (equation (4)).

Two very clear trends can be seen in figure 5. First, for each
fixed p the average species richness of reshuffled communities at first
increases as the number of colonizers increases, and then levels off
and remains essentially constant over a large range of numbers of colonizers. Second, the "saturation value" of reshuffled species richness
is a monotonically decreasing function of the probability p.

Thus, the species richness of these communities is very strongly
controlled by dominance relationships within the pool of species available for colonization. (The number of colonizing species is less important!) Some such idea is implicit in much of the literature on sessile communities (for instance, see the remarks of Poore 1964), but it
emerges with striking clarity here.

Of course, in natural communities niche differentiation will
allow, to a greater or lesser extent, for more species richness; a
complete solution to the species packing problem for communities where
spatial competition is important must involve a combination of dominance ideas and niche ideas. The point is: neither dominance ideas nor
niche ideas alone will suffice.

*

This author is unaware of any data from the field which directly
test the behavior shown in figure 5. However, the marked tendancy of
species-area curves from temperate zone terrestrial plant communities
to level off at high areas (for instance, Hopkins 1955) could be due
to this "saturation" property of reshuffles.

To see this, we reason as follows. Consider first a founder controlled community. Its species-area curve should increase monotonically with, typically, the number of species per quadrat proportional
to the logarithm of quadrat area at large areas (Pielou 1969, p. 218;
May 1975, p. 103). Now, in a dominance controlled community, if one
could sample the number of species which had colonized each quadrat
one would obtain this same behavior. But what we actually measure is
the number of species remaining in these samples after a reshuffle, and
this will eventually level off as quadrat area (hence the number of
colonizing species) becomes large enough, because of the "saturation"
behavior shown in figure 5.

Another area effect can also be understood in terms of these
ideas. Jackson (1977b) followed the development of epibenthic communities on asbestos-cement panels of different sizes submerged in Dis-
covery Bay, Jamaica. He found that while species richness was larger
on the larger panels (with some of the smallest panels being monopolized by a single species, suggesting that these panels were of the
order of the appropriate patch size for these communities), species
composition was more predictable on the larger panels. That is, the
outcome of a reshuffle is predictable only if the area is large enough
to accomodate a large enough sample of colonizers. Species which would
be eliminated in the reshuffle in a large area may be able to monopo-
lize a smaller area before other species are able to colonize. Jackson suggests this as a possible mechanism enabling some species to
persist in the relatively small, cryptic habitat islands provided by
the skeletal undersurfaces of foliaceous corals such as Agaricia, Montastrea, and Mycetophyllia in Caribbean reefs.

*

Unfortunately, it proved too cumbersome to determine the species

Figure III.6. Some typical averaged abundance-rank curves for reshuffled communities, calculated as explained in the text.

richness of the communities with quasicycles. However, recall that a quasicycle involves species which, though dominated in some patches, are able to invade and predominate for a time in other patches rather than being eliminated from the system. Thus the systems with quasicycles will in general be richer in species than corresponding systems without quasicycles. Jackson and Buss (1975) suggest that this may help explain the high species richness of some tropical systems.

III.4.3. Numerical dominance.

A further quantity was calculated for those communities which reached an equilibrium in the reshuffle simulations. For each number n of colonizing species, probability p, and final number of species, the abundance (in terms of number, biomass, or whatever other units may be appropriate for the densities) of each species in the community was calculated, and these quantities were ordered from largest to smallest and then averaged for each rank.

This enables us to examine numerical dominance relations for the unperturbed equilibria of our reshuffled communities. Because the perturbations to the equilibrium densities are again small (Section 4.4), this gives a good approximation to the numerical dominance relations for the perturbed equilibria.

Several of the resulting abundance-rank curves are plotted in figure 6. The reader who is familiar with such curves will recognize that the steep, species-poor curve is more or less of the form for a geometric series, while the curves from richer communities are roughly lognormal in form.

While there is no lack of exceptions, this trend seems to be roughly followed in terrestrial plant communities. Typical examples can be seen in figures 7 and 8, which depict actual abundance-rank curves from natural communities.

*

Let p_i be the proportion of "abundance" (in whatever units are appropriate) in the ith species. Then the Simpson (1949) index for "concentration of dominance" (Whittaker 1965) is given by

$$C = \sum_i p_i^2.$$

This is simply related to the probability that two individuals chosen from the community at random belong to the same species (this probability being $(NC-1)/(N-1)$, where N is the total number of individuals in the community), and is particularly sensitive to a concentration of abundance in one or two species.

It can be converted to an evenness index by writing

(III.11) $E = 1/SC$,

where S is the number of species in the community. The index E, which is sometimes called the Simpson-Horn evenness index, takes on values between 0 and 1, with a higher value meaning that abundance is more evenly apportioned among the species in the community (less dominance). There are several other evenness indices (May 1975, Pielou 1975), but

63

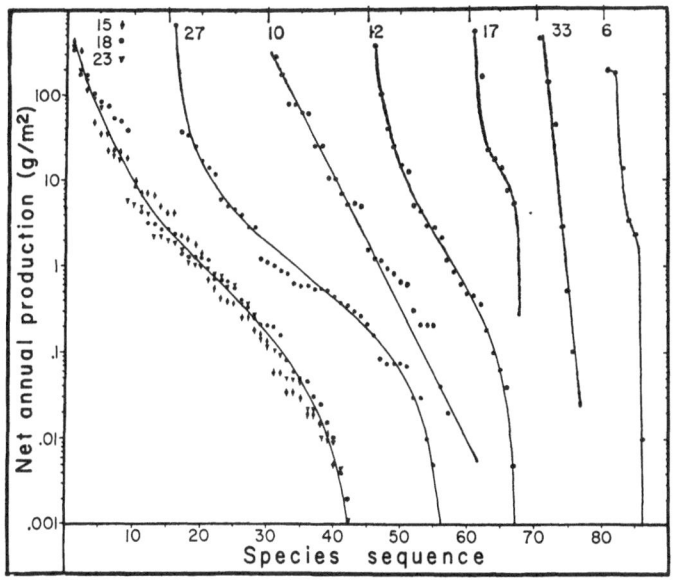

Figure III.7. Abundance-rank curves for vascu-
lar plant communities in the Great Smoky Mountains, USA. (From Whittaker 1965. Copyright
1965 by the American Association for the Advance-
ment of Science.)

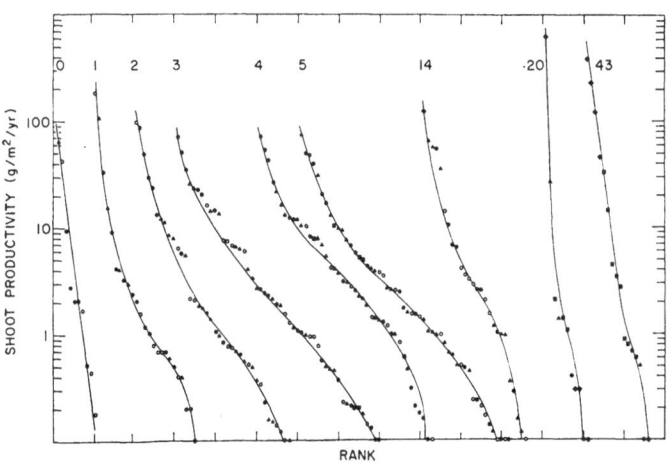

Figure III.8. Abundance-rank curves for a suc-
cessional sequence in an abandoned old field near
Brookhaven National Laboratory, New York, USA.
(From Whittaker 1972.)

they usually give similar results to this one.

It appears that evenness and species richness are positively correlated in a wide variety of sessile communities in nature, including grasslands (Singh and Misra 1969; Krishnamuthy 1977), forests (Risser and Rice 1971), and coral reefs (Porter 1972). The communities involved in each of these studies were all at comparable successional stages; studies which include communities at different successional stages tend to find that although there are correlations at comparable stages of succession, evenness and richness are not correlated throughout succession (Shafi and Yarranton 1973; Nicholson and Monk 1974; Tramer 1975).

In order to test for a correlation in our theoretical reshuffled communities, 100 communities were randomly constructed as follows. First, for each community a positive interval of real numbers was randomly chosen, and the competition coefficients for the community were selected at random from this interval. These intervals were chosen subject to the constraint that the probability p (equation (4)) was always in the range (0.01,0.1). Then for each community colonization and reshuffle were simulated as in Section 3, and the Simpson-Horn index E and species richness S of the reshuffled communities which reached equilibrium were calculated. Finally, the correlation coefficient for these two quantities was calculated. This showed a very highly significant correlation between evenness and richness (r=.52, P=.00000003).

III.4.4. Stability.

The parameter Λ, which characterizes the local stability of an equilibrium (see Section II.2), is calculated for reshuffled equilibria in Section A.3.2. In that section its average value over certain ensembles of randomly chosen systems is also calculated. The significance of these computations is much the same as in the case of founder control (Section II.2).

Firstly, while there is some dependence of stability on species richness of the reshuffled communities, this dependence is again so highly sensitive to the value of the dispersal parameter D that it will in practice be masked by random effects. Therefore, in effect: for each D these equilibria are either stable for all S>1, or for none;

species richness is determined by the reshuffle process rather than by local stability.

Secondly, we find from table A.1 that the values of D at which these equilibria become unstable are again very small. This justifies the use of first-order perturbation calculations for these equilibria, in particular to establish positivity of the equilibrium densities in Section 4.1.

III.4.5. Total standing crop.

In Section II.3 we calculated the relation between average total standing crop (that is, average total biomass in the community) and species richness for founder controlled communities. This should also be a good approximation to the average total standing crop for dominance controlled communities. We examine the qualitative behavior of this quantity here.

We can roughly approximate the abundance-rank relations in our reshuffled communities (Section 4.3) by the geometric series which result from colonization by space pre-emption in Chapter II. This case was treated in Section II.3, where it was found that equation (II.7) together with the substitution (II.8) gives the average total standing crop as a function of the parameters e, f, g, h, m, k, D specifying the ensemble of systems over which the total standing crop has been averaged.

Choosing e, f, g, and h as in the simulations described in earlier sections of this chapter, we see that the factor involving g and h in (II.7) is always the same, and that the factor in square brackets in (II.7) increases as reshuffled species richness increases (using figure 5). We have already remarked in Section II.3 that the factor $1-k/(2-k)$ increases as species richness increases. Therefore, according to (II.7) and (II.8), average total standing crop should decrease as species richness increases.

However, several field studies have found a positive correlation between total biomass and species richness in sessile communities: Singh and Misra (1969) observed this in grasslands, Risser and Rice (1971) in forests, and Kohn and Leviten (1976) for intertidal gastropods.

This positive correlation is probably due to environmental heterogeneity. The negative correlation predicted by our calculation is a very small effect (first order in D), so it could easily be masked by other factors which influence biomass. Indeed, Kohn and Leviten (1976) find that both total biomass and species richness are positively correlated with habitat structural complexity for their sedentary intertidal gastropods, so at least in this case there is certainly an element of environmental heterogeneity involved —— and our model explicitly assumes a lack of such heterogeneity (Section I.4, item 1a)).

III.5. Invasion and succession.

In Section I.5 we assumed that the flow of biomass of any species into the area occupied by one of our communities is negligible compared with the flow of biomass of that species within the community due to interpatch dispersal. This has enabled us to treat these systems as biotically closed, except for an initial colonization of empty space, to which this assumption obviously does not apply (Section I.5).

There is another circumstance under which this assumption breaks down. The assumption is very reasonable when applied to species which are already established in the community. But suppose there is a flow of biomass into an established community of some species which is itself not predominant in any patch of the area. Then even though this may be a very small flow, there is no other flow compared with which it can reasonably be neglected; it is, as in the case of colonization of empty space, the entire population dynamics of the new species and therefore cannot be ignored. The new species may be able to _invade_ the community.

We have already done most of the mathematical thinking that is needed to understand this phenomenon. The requirements for a potential invader from outside the community to become established in some patch are precisely the same as those for a species trying to invade that patch from another patch within the community; these requirements have been discussed in Section 2. Once an invading species has become predominant in any one patch of the area, there will be a new reshuffle, following the same rules as the original reshuffle among the colonizers, until a new community (or a new stage in the development of the community) is established. Of course, like the original colonization

process, the process of invasion will often be mixed together temporally with reshuffles.

As additional species invade the community, its character will
gradually change. This time evolution of the species composition of
a community is known as <u>succession</u>.

The idea of succession has long (see, for instance, Clements
1916) provoked ambitious theoretical generalizations, but the reality
of the phenomenon has proved too complicated to be encompassed by any
one scheme. It seems that succession must be understood rather by the
drawing of precise, appropriate distinctions. While such distinctions
have, of course, been implicit in the thinking of many students of
succession, Horn (1976) has articulated them with particular clarity.
We will, then, conduct our discussion within the framework established
by Horn.

He distinguishes three successional modes:

1) <u>Chronic</u>, <u>patchy</u> <u>disturbance</u>. In this case, the system is subject to frequent, more or less localized (in time and in space) disturbances, which serve to clear patches of space for fresh colonization
(primarily from the surrounding community). The most thoroughly studied example is the rocky intertidal zone, where the disturbances come
from damage by waves and floating driftwood (Dayton 1971). Some observers, for instance Lloyd, Inger, and King (1968), have suggested that
damage from intense rains may play an important role in maintaining
diversity in tropical forests, which may put some of these forests in
the present category. This view has been very cogently articulated
by Connell (1978).

If the time scale for disturbance is sufficiently short relative
to the time scale for reshuffle, competitive relationships may not be
important in structuring the community; then models like ours are not
very appropriate (see also Section IV.5). Appropriate models have
been developed by Levin and Paine (1974, 1975). However, at intermediate disturbance rates our treatment of harvesting in Chapter IV might
possibly be a reasonable approximation.

2) <u>Obligatory</u> <u>succession</u>. This is characterized by the necessity
of early successional species to pave the way for later species: later
species must await the dominance and decline of earlier ones. (It is
a non-cyclic counterpart to the Watt cycles discussed in Section 3.)

Classic examples (Kershaw 1973, Ch. 3) are provided by communities on sandy or boggy soils: here accumulation of organic matter from appropriately adapted early successional plant forms increases the water retention of sandy soil or the depth and drainage of peat, modifying the soil in favor of plants which are adapted to intermediate degrees of moisture; such plants then colonize.

Succession in terrestrial plant communities often takes the form of a sequence of wave-like invasions by strata (for instance, Drew 1942; Oosting 1942; Bazzaz 1968; Reiners, Worley, and Lawrence 1971; Nicholson and Monk 1974). This does not necessarily indicate an obligatory succession (see the third category, below), but probably often does.

As we have pointed out in Section I.4, the models studied in this volume do not encompass the sort of environmental changes involved in obligatory succession. However, our models should be applicable to the development of a single successional stage.

3) _Competitive_ _hierarchy_. In many north temperate forest successions, late successional species are present (but not yet numerically dominant) almost as soon as early successional species invade an abandoned field (for instance, Hope-Simpson 1940; Drury and Nisbet 1973). Thus the late successional species are late not because they can invade only under conditions which are established by the earlier species (as in obligatory succession), but because their ability to dominate becomes relevant only after a certain level of crowding has been reached.

Our models should apply to this kind of succession.

Horn does not suggest that all successions can be uniquely partitioned into these three categories, but rather that most successions reflect some combination of these three processes, with one mode of succession often more prominant than the others.

Applying our models to successions which are governed by competitive hierarchy, or to single stages of obligatory successions, would give equivalent results to the phenomenological Markov chain models which have already been extensively applied to various specific examples. This literature has recently been reviewed by Horn (1975, 1976).

We mention here just one important trend which follows immediately from the notion of succession as a process of invasion and reshuffle. The final stage in this process will be, after all potential invaders have had a chance to become established, a final reshuffle. In this final reshuffle, species richness will decrease as some species are eliminated (recall Section 4.2).

Declining species richness in late succession has been documented in a variety of terrestrial plant communities (for instance, by Odum 1963; Heilman 1966, 1968; Viereck 1966; Loucks 1970; Auclair and Goff 1971; Nicholson and Scott 1972; Shafi and Yarranton 1973). Many of these authors have also noticed a tendency for the decline in richness to be accompanied by a decline in evenness, in agreement with the extrapolation of our results in Section 4.3 to succession. There have also been reports (Oosting 1942; Quarterman 1957; Odum 1960; Nicholson and Monk 1974) of declining species richness toward the end of individual obligatory successional stages.

While some authors (Whittaker 1965; Monk 1967) have seen indications of a continuing increase in richness in late succession, it appears that decreasing richness is the more typical behavior (lending, incidentally, support to the notion that periodic devastation by fire may help maintain forest diversity, as argued by Loucks 1970; Day 1972; Taylor 1973; Heinselman 1973; Habeck and Mutch 1973, and discussed further in Section IV.5).

However, these studies were all conducted in the temperate and arctic zones; quite different processes may be at work in the tropics, as proposed in Section V.3.

III.6. Other forms of dispersal.

We have in this chapter restricted our attention to the extreme case of maximum interpatch mobility which we call random dispersal (equation (I.7)). However, the qualitative behavior that we have found obviously does not depend in any essential way on this restriction; other forms of dispersal will produce much the same qualitative behavior.

There is, however, one aspect of community structure for which the details of dispersal clearly will play an important role: this is the phenomenon of spatial clumping, which we have briefly alluded to in Section I.1.

This patchiness has many causes, but one of these appears to be the low spatial mobility of some species. For example, some plant species which occur in unusually clumped spatial patterns have particularly heavy seeds, which therefore spread less readily than the seeds of other species (Ashby 1948). Other species, such as Sassafras albidum and Diospyros virginiana, occur in clumps because they sprout from root fragments in old fields (Drew 1942; Potzger and Potzger 1950).

This mechanism can also be seen in marine intertidal communities. Thus the alga Postelsia palmaeformis occurs very patchily within beds of Mytilus californianus, largely because it has a very limited spore dispersal distance (Dayton 1973a).

III.7. Summary.

This chapter treats communities which are formed from colonizing species for which the competition coefficients α_{ij} (i≠j) are randomly chosen from a real interval (e,f) with $0<e<1<f$ and $p=(1-e)/(f-e)<0.1$.

Two concepts of interspecific dominance, namely numerical dominance and functional dominance, are defined and discussed in Section 1.

In Section 2 it is argued that in these communities there will be after colonization a "reshuffle" of species among the patches, with some species invading many patches (thereby establishing the numerical dominance so characteristic of sessile communities in nature) and some species being eliminated from the community altogether. The rules governing such a reshuffle are derived from the dynamical equations, and it is pointed out that a reshuffle can develop in two ways: either it terminates after a finite number of interpatch invasions, or it does not.

Section 3 discusses the case in which the reshuffle never terminates. It is argued that while the sequence of invasions experienced by at least some patches will in this case be cyclic, the time evolution of the system as a whole will not necessarily be cyclic, but may instead fill out a toroidal "strange attractor" in phase space. This behavior is termed a "quasicycle". The probability for the occurrence of quasicycles, as a function of the probability p and the number of colonizing species, is calculated, and it is found that quasicycles may be quite likely for systems with a rich supply of colonizing species. However, biological arguments are presented which suggest that this is probably an overestimate of the likelihood of quasicycles.

Holling's ideas about the "resilience" of complex systems are briefly discussed and related to the present work. The cyclic vegetational changes first observed by Watt (1947) are contrasted with quasicycles.

The section closes with the observation that in some cases random influences will produce instead of a quasicycle a random sequence of interpatch invasions; this behavior is termed a "lottery". It may conceivably have been observed by Sale (1975) among territorial coral reef fish.

Section 4 treats the case in which the reshuffle terminates after a finite number of interpatch invasions and an equilibrium state is reached. Observations by Dayton (1973b) which illustrate this phenomenon in an intertidal zone are portrayed. In Section 4.1 it is shown that in our models the equilibrium densities are positive.

In Section 4.2 average equilibrium species richness is calculated as a function of the probability p and the number of colonizing species. It is found that species richness depends more on p than on the number of colonizing species; thus it is determined primarily by interspecies (functional) dominance relations. It is argued that the levelling off at large areas of species-area curves from many terrestrial plant communities, as well as another area effect observed by Jackson (1977b), can be understood in terms of these ideas.

In Section 4.3 abundance-rank curves for the equilibrium communities are calculated and compared with observations. Data showing that evenness and species richness are often positively correlated in natural sessile communities are cited, and it is found that these quantities are highly correlated in our theoretical equilibrium communities.

Local stability of the equilibria is discussed in Section 4.4, with reference to detailed calculations in Appendix A. It is argued that local stability has no significant effect on species richness, and that first-order perturbation theory is an appropriate mathematical tool when the equilibria are stable.

In Section 4.5 our only conflict with observational data is found. It is shown that in these models the average total standing crop at equilibrium is inversely related to species richness, and field data contradicting this are cited. It is suggested that this discrepancy may be due to environmental heterogeneity in the field.

Invasions of an established community from outside are discussed in Section 5, and succession is depicted as a process of invasion followed by new reshuffles. It is pointed out that this conception implies that species richness should decline in late succession, and toward the end of individual obligatory successional stages, and observational data are surveyed.

Section 6 briefly sketches the role of differential ranges of dispersal in producing clumped spatial distributions. Examples from the field are cited.

Chapter IV. HARVESTING

We consider in this chapter the influence of harvesting —— used
here to mean any more or less continuous (in space and in time) removal
of biomass from a community, be it by grazing, by predation, or by intentional or unintentional human action —— on the competitive communities studied in the previous two chapters. Few of the corresponding
natural communities are free of harvesting, and it is in fact often a
major determinant of community structure.

For instance, harvesting can have a pronounced effect on diversity, and we will emphasize this effect in our discussion, for there
is a rich natural historical literature (which we review in Section
4) devoted to it.

IV.1. Harvesting.

It is well known that the regular removal of biomass from competitive systems profoundly affects competitive relationships. For example, Darwin (1859, Ch. 3) allowed a lawn which had long been mown
to grow freely, and found that "out of twenty species growing on a
little plot of mown turf (three feet by four) nine species perished,
from the other species being allowed to grow up freely". In another
classic study, Paine (1966) found that removal of the top predator
from an intertidal community resulted in a decrease in the number of
major space utilizing species in the community from fifteen to eight
(see also Paine 1974).

We can treat such effects by adding a term to equation (I.6), so
that

$$(\text{IV.1}) \qquad \frac{dx_i^\mu}{dt} = C_i x_i^\mu (1 - \sum_j \alpha_{ij} x_j^\mu) + D \sum_\nu E_i^{\mu\nu} x_i^\nu - H_i^\mu(x) \qquad \begin{array}{l} \mu = 1, \ldots, m \\ i = 1, \ldots, n \end{array}$$

where each <u>harvesting rate</u> H_i^μ is, in general, a function of all $m \times n$ x_j^ν.

This sort of model is most directly appropriate to a situation in which biomass is extracted by some agent thought of as external to the system (hence the terminology: harvesting), but it can also be thought of as a rough approximation to grazing or predation, or as the effect of some environmental degradation or disturbance which acts to decrease the net growth rates (for instance, by increasing the death rates or by decreasing the birth rates).

Again, we have omitted some biological detail in the hope of arriving at a fairly general understanding of harvesting. In the case of grazing or predation, we omit the dynamics of the grazers or predators altogether, possibly missing important feedback effects between trophic levels. One such effect has been considered by Caswell (1978), whose conclusions, however, do not conflict with ours. Disturbances will often be so irregular (in space or in time or both) that they cannot sensibly be modelled by a deterministic function of the densities (see, for instance, the remarks of Levin and Paine 1974, 1975; and the further comments in Section 6 below). Age effects are also sometimes important (Vandermeer 1975).

On the other hand, the effects of all these forms of harvesting have been observed to be remarkably similar, which suggests that it does make sense to ignore a good deal of detail.

We shall assume here that each harvesting rate H_i^μ depends only on the <u>local</u> density of species i, and that this dependence is the same in every patch:

$$H_i(x_1^1, x_2^1, \ldots, x_n^m) = H_i(x_i^\mu).$$

This is a reasonably typical behavior, but it is certainly not the only realistic possibility. It ignores, for example, the need of specialized grazers or predators to search out their preferred food. It also does not allow for the "switching" behavior of some predators,

which develop a disproportionate preference for the globally most
abundant prey species at any time (see, for instance, Murdoch 1969).
Modelling this sort of behavior would require a dependence of the
harvesting rates also on the quantities $\sum\limits_{\nu} x_j^{\nu}$.

Specifically, we shall take each H_i to have the form of Holling's
"Type I functional response":

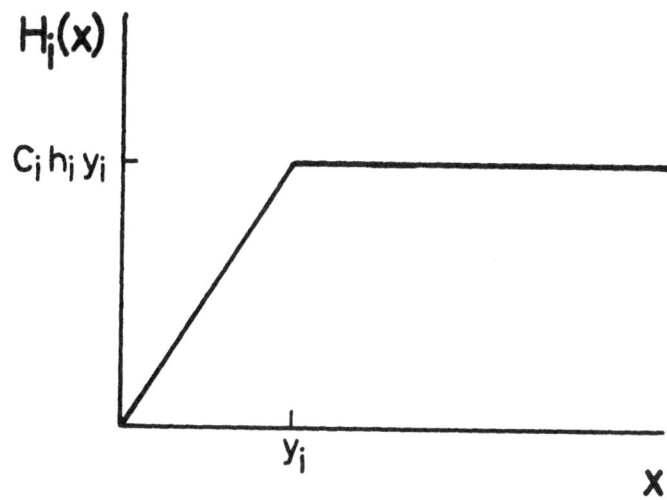

$$(IV.2) \qquad H_i(x) = C_i h_i \times \begin{cases} x \ , & x \leqslant y_i \\ \\ y_i, & x > y_i \ , \end{cases}$$

where C_i is the same C_i that occurs in the Lotka-Volterra local dynamics (I.4), and h_i and y_i, i=1,...,n, are also positive constants.
This is the simplest form which embodies the two essential features
of most harvesting rates: 1) increasing removal of biomass with increasing abundance of the harvested population (in our case, constant biomass removed/biomass present) at low densities until 2) a saturation
level of harvesting is reached, after which a constant amount of biomass is removed. For other forms, see, for instance, Ivlev (1961) or
Holling (1965).

The dynamics of one species i, in the absence of all competitors and without spatial dispersal, is then given by

$$\frac{dx_i}{dt} = C_i x_i (k_i - x_i) - H_i (x_i).$$

Suppose the parameters are chosen so that such a one-species system has a stable equilibrium at $x_i = x_{ie}$. It will be convenient to introduce the dimensionless parameters κ_i, ζ_i by

(IV.3) $x_{ie} = \kappa_i k_i$, $h_i = \zeta_i k_i$.

There are four possible structurally stable qualitative structures for these one-dimensional systems. We shall refer to them as harvesting structures. The corresponding qualitative forms of the growth rates dx_i/dt are depicted in figures 1-4. (The transitional case $1 - \zeta_i = 1 - \kappa_i$ is structurally unstable; this is of no significance.)

In the case of structure S1, the harvested species is not sufficiently abundant to support the saturation level of harvesting; the harvester is food-limited. In this case our neglect of the harvester dynamics is questionable. In structures S2, S3, and S4 the harvester is limited by factors other than food, and our neglect of the harvester dynamics is probably quite harmless.

Several large mammalian herbivores in the African savanna have been shown to be food-limited (Sinclair 1974; Phillipson 1975; Coe, Cumming, and Phillipson 1976), but Hairston, Smith, and Slobodkin (1960), in their famous "green Earth" argument, maintain that most herbivores tend to be limited by factors other than food, such as predation and disease. Despite criticism of their argument by Murdoch (1966) and by Ehrlich and Birch (1967), and indeed the existence of some known exceptions, it seems widely accepted as a rough generalization that there is a tendency for herbivores to be limited by other factors than food.

For another organism that will be of interest to us, the predatory starfish Acanthaster planci (the "crown of thorns" starfish),

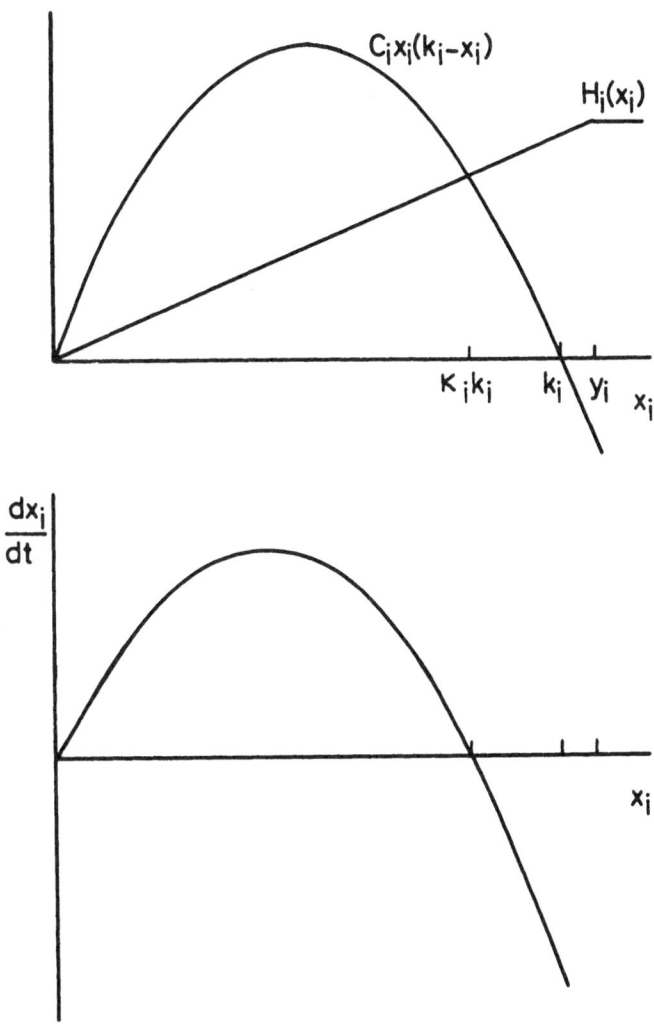

Figure IV.1. Harvesting structure S1: $1-\zeta_i = \kappa_i$.

there is spectacular evidence of apparent nonfood-limitation under normal circumstances. During the last ten years, a number of _Acan-thaster_ populations in the Indo-Pacific region, particularly on the Great Barrier Reef (Australia), have reached "plague" proportions. The causes of these population explosions are still not understood

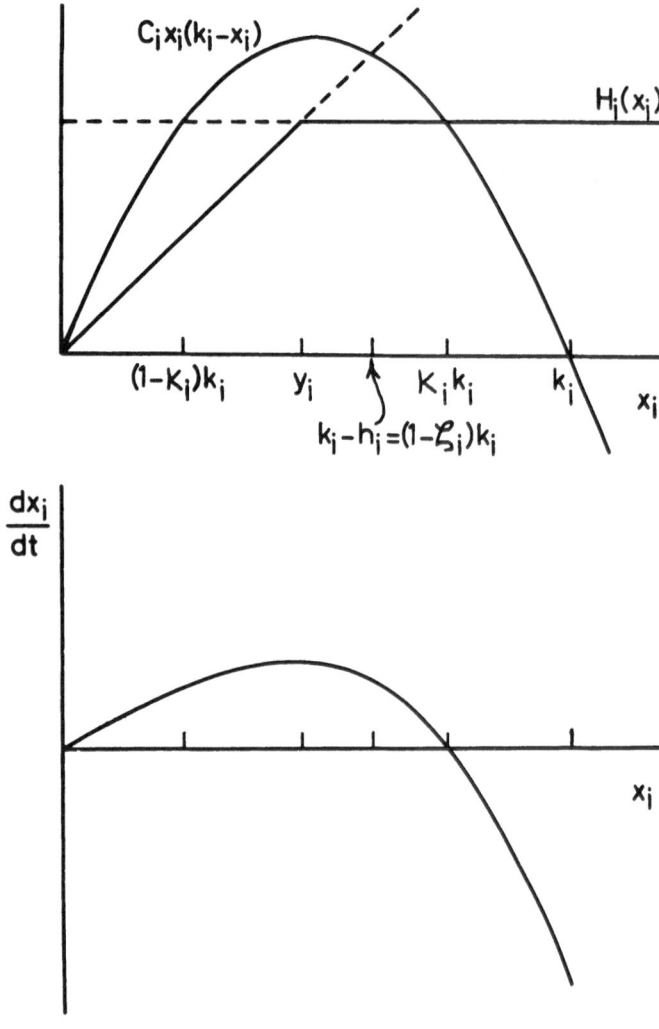

Figure IV.2. Harvesting structure S2: $1-\kappa_i < 1-\zeta_i < \kappa_i$.

(for a recent review, see the report of Sale, Potts, and Frankel 1976), but it seems unlikely they can be explained in terms of food-limitation. Indeed, Glynn (1973) estimates that Panamanian reefs could still maintain positive growth if <u>Acanthaster</u> populations there were to increase to 2.5 times their present density.

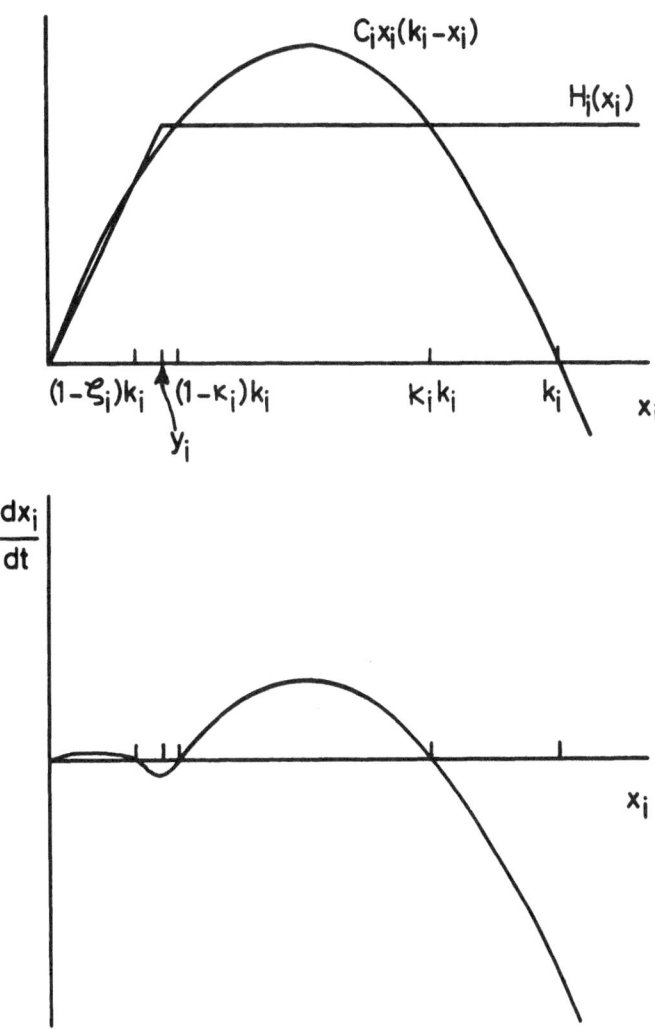

Figure IV.3. Harvesting structure S3: $0 < 1-\zeta_i < 1-\kappa_i$.

In the other extreme case, structure S4, the harvester creates an Allee-type effect: the net growth rate of the harvested population is negative at small densities. The harvested population will, then, be driven to extinction (at least locally) if for some reason its density falls too low.

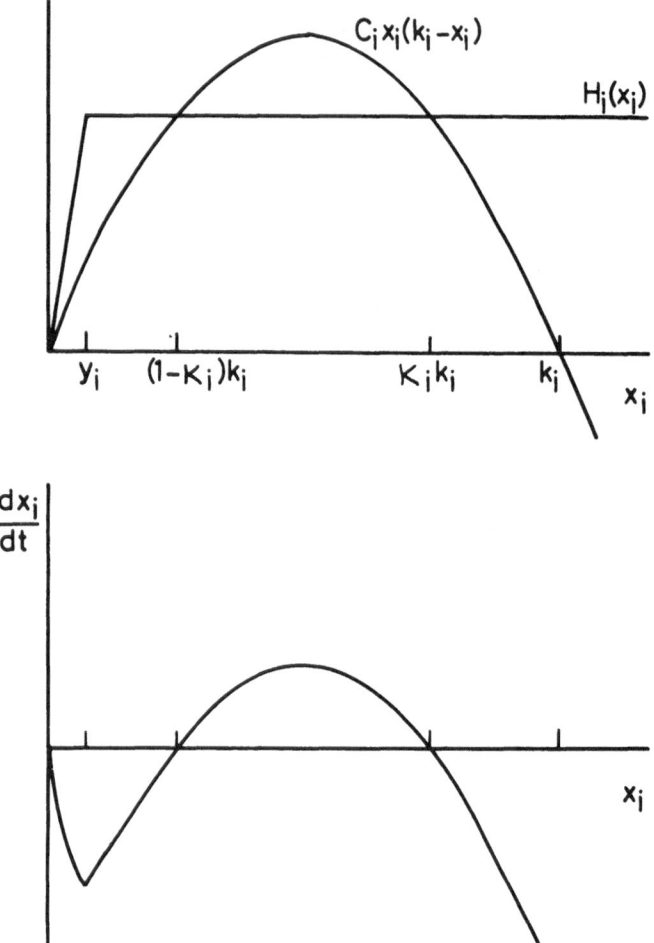

Figure IV.4. Harvesting structure S4: $1-\zeta_i < 0$.

This sort of feeding behavior is by no means uncommon. Evidence of it can be seen, for instance, in grazing by large mammals in forests (Peterkin and Tubbs 1965) and by limpets (Jones 1948), fish (Randall 1965), and sea urchins (Kain and Jones 1966; Leighton, Jones, and North 1966; Dayton 1975b)on marine algae, as well as in predation by snails

(Fisher-Piette 1935; Connell 1961b, 1970a; Luckens 1970; Dayton 1971) and by starfish (Paine 1966, 1971; Dayton 1971) in intertidal zones.

This variety of possible harvesting structures has some very interesting implications for the qualitative behavior even of these simple single-species models. This topic has recently been discussed by May (1977).

*

It will be convenient to have an expression for the _yield_ at equilibrium,

$$Y_i = H_i(\kappa_i k_i),$$

in terms of the parameters κ_i and ζ_i. For the case of food-limitation (structure S1), this is simply

(IV.4) $\qquad Y_i = C_i \kappa_i \zeta_i k_i^2 = C_i \kappa_i (1 - \kappa_i) k_i^2.$

For nonfood-limitation (harvesting structures S2, S3, S4), we first calculate the saturation density y_i as

(IV.5) $\qquad y_i = \kappa_i (1 - \kappa_i) k_i / \zeta_i;$

then the yield at equilibrium is

(IV.6) $\qquad Y_i = C_i \kappa_i (1 - \kappa_i) k_i^2.$

The upper bound on the yield at equilibrium such that there is a stable equilibrium at all is called the _maximum sustainable yield_. This is

(IV.7) $\qquad Y_{MSi} = C_i k_i^2 / 4$

for all four harvesting structures.

*

We have in this section allowed arbitrary units of measure for the densities, in order that the dimensional nature of the various parameters could be made clear. Henceforth we return to our convention (Section I.3) of expressing the densities in units such that all carrying capacities k_i are equal to 1.

IV.2. The reshuffle.

We shall study once again communities which arise from the colonization of empty space (as described in Section I.5), but now we shall take into account harvesting by using equation (1) to describe the dynamics. We expect that in general there will be a reshuffle (Section III.2) of species among the patches, which will result in an equilibrium, a quasicycle, or a lottery. However, the rules governing such a reshuffle will now be somewhat different from those derived in Section III.2 for systems without harvesting.

Our object in this section is to find these rules. In order to do so, we proceed just as we did in Section III.2. Rather than repeat the entire argument here, we shall only list the changes to the discussion in Section III.2 that have to be made for our present purposes. It may be well for the reader to refresh his memory of that section at this point.

1. Suppose that species i attempts to invade a patch which is predominated by species s. Then while the density of species i is small, its growth rate is to first order in D

$$\frac{dx_i^\mu}{dt} \simeq C_i x_i^\mu (1-\zeta_i-\alpha_{is}\kappa_s) + D \sum_{\nu:c(\nu)=i} x_i^\nu.$$

Therefore, the set of potential invaders of a patch with one predominant species s is now

(IV.8) $I(s) = \{i| \ 1-\zeta_i-\alpha_{is}\kappa_s>0\}.$

Similarly, for a patch with two predominant species s and t, with densities x_s and x_t, the set of potential invaders is

$$I(s,t) = \{i| \ 1-\zeta_i-\alpha_{is}x_s-\alpha_{it}x_t>0\}.$$

Of course, x_s and x_t are now <u>not</u> equal to the Lotka-Volterra two-species equilibrium values (III.6): harvesting will alter these equilibrium densities.

2. In accordance with these changes, it is now necessary to alter Hypothesis III.1 to read

<u>Hypothesis IV.1a</u>. If the set $I(s)$ of potential invaders of some patch in which species s is predominant is nonempty, then that potential invader i for which $1-\zeta_i-\alpha_{is}\kappa_s$ is the largest, and only that potential invader, will successfully invade the patch.

<u>Hypothesis IV.1b</u>. If the set $I(s,t)$ of potential invaders of some patch in which species s and t are predominant is nonempty, then that potential invader i for which $1-\zeta_i-\alpha_{is}x_s-\alpha_{it}x_t$ is the largest, and only that potential invader, will successfully invade the patch.

3. The outcome of an invasion can be affected by harvesting.

The outcome of an invasion of a patch with one predominant species is determined by the within-patch dynamics of the corresponding two-species system. But the qualitative structure of such a system, as expressed by the topology of its phase space trajectories, is, in general, altered by harvesting. Thus in order to find the outcomes of interpatch invasions, we have to study these structural changes induced by harvesting. This is done in Appendix C, and the relevant results are summarized in Theorems C.1 and C.2.

For a patch with two predominant species it is much more difficult to determine the result of an attempted invasion. However, it is eminently reasonable to suppose that stable three-species equilibria are still extremely rare (recall the discussion of figure III.2). Therefore, we assume that if there is a stable equilibrium with less than three species, the system will approach this equilibrium.

Our exact procedure is as follows. Suppose that species i invades a patch which is predominated by species s and t. We first search the i-s and i-t planes for equilibria which are stable as equilibria of the three-dimensional system consisting of species i, s, and t. In case either of these planes contains more than one stable equilibrium, we use Theorem C.1 to find the appropriate one.

Now, if only one of these planes contains such an equilibrium, we assume that the system approaches it asymptotically. If both planes contain such an equilibrium, we choose one of them at random (this corresponds to the randomness in the quantities ε_1 and ε_2 in a real system) and assume that it is approached asymptotically. If neither plane contains a stable equilibrium, we check whether the equilibrium at $(x_i, x_s, x_t) = (\kappa_i, 0, 0)$ is stable as an equilibrium of the three-dimensional system. If so, we assume that the system approaches this equilibrium asymptotically. If not, we throw the system out of our statistical sample, as one in which three species will predominate in some patch (this occurs only rarely in our calculations, however).

IV.3. Influence of harvesting: theoretical.

We consider in this section the influence of harvesting, as expressed in equation (1), on community structure. We will treat "indiscriminate harvesting", in which all species are harvested at roughly comparable rates, the most thoroughly, for this is intuitively the least obvious case, and it is important as a paradigm for general predation pressure from higher trophic levels. We will also discuss some important forms of selective harvesting.

There have been several demonstrations (Slobodkin 1961; Parrish and Saila 1970; Cramer and May 1972; Slatkin 1974; Roughgarden and Feldman 1975; Yodzis 1976a; Caswell 1978) that harvesting can, under a variety of circumstances, make possible the coexistence of two species which cannot coexist without harvesting. This observation will be amplified and sharpened here for those cases where the harvested species compete for space. This includes the vast majority of documented examples from the field (Section 4). Also, the discussion here will not be limited to the case of two harvested species; the number of harvested species will be arbitrary.

We can reach a good general understanding of the influence of harvesting on these communities by looking at how it affects the probability p for a randomly chosen species to be a potential invader of a patch predominated by another randomly chosen species. (Recall from Chapter III that many aspects of community structure are determined largely by the value of p.)

From (III.3) and from equation (8) of this chapter, we see that the condition for species i to be a potential invader of a patch predominated by species s can be expressed as an upper bound on the competition coefficient α_{is}:

$$\alpha_{is} < B_{is}$$

for some positive number B_{is}. Therefore, if the competition coefficients are randomly chosen from an interval (e,f) independently and with uniform probability distribution, the probability for species i to be a potential invader of a patch predominated by species s is

$$(IV.9) \qquad p_{is} = \begin{cases} 0 & , \ B_{is} \leq e \\[2ex] \dfrac{B_{is}-e}{f-e} & , \ B_{is} > e. \end{cases}$$

The upper bound B_{is} is given by (III.3) and (8) as

$$(IV.10) \qquad B_{is} = \begin{cases} 1 & \text{without harvesting} \\[2ex] \dfrac{1-\zeta_i}{\kappa_s} \cdot & \text{with harvesting.} \end{cases}$$

The domain of κ_s is ($\sqrt{2}$,1) and the domain of ζ_i is (0,∞). Therefore, the domain of B_{is} when there is harvesting is ($-\infty$,2).

We see, then, from (9) that the effect of harvesting will usually be to cause the probability p_{is} to decrease; only under special circumstances will it cause p_{is} to increase. Therefore we expect on the basis of the calculations in Chapter III that, for example, harvesting will usually make it possible for more species to coexist in our communities. This observation will be sharpened in the remainder of this section.

*

From an empirical point of view, the most directly meaningful quantities for specifying a harvesting rate of the form (2) are the saturation density y_i and the yield Y_i (equations (4) and (6)). The parameters κ_i and ζ_i are thereby determined as

$$(IV.11) \quad \kappa_i = (1+\sqrt{1-Y_i/Y_{MSi}})/2$$

$$(IV.12) \quad \zeta_i = \begin{cases} (1-\sqrt{1-Y_i/Y_{MSi}})/2 & \text{for structure S1} \\ \dfrac{Y_i/Y_{MSi}}{4y_i} & \text{for structures S2, S3, S4,} \end{cases}$$

where Y_{MSi} is the maximum sustainable yield (7).

We proceed to analyze some important special cases in more detail.

IV.3.1. Indiscriminate harvesting.

We assume in this case that all species are harvested at roughly comparable rates. In addition to describing a single harvester which adopts this particular harvesting policy, indiscriminate harvesting provides a rough approximation to general predation pressure from higher trophic levels, involving many harvesters, or to the effect of general environmental degradation such as pollution with toxic substances.

In the idealization in which all species are harvested in exactly the same way, we have $y_i = y = $constant and $Y_i/Y_{MSi} = Y/Y_{MS} = $constant for all i, whence via (11) and (12) $\kappa_i = \kappa = $constant and $\zeta_i = \zeta = $constant for all i. It can easily be seen from figures 1-4 that we will always have for each i: $1-\zeta_i < \kappa_i$. Therefore it will always be the case for indiscriminate harvesting that

(IV.13) $\quad \dfrac{1-\zeta}{\kappa} < 1.$

Hence the probabilities p_{is} given by equation (9) can never increase under indiscriminate harvesting.

Consider first indiscriminate harvesting of a founder controlled community. Without any harvesting all the invasion probabilities p_{is} are zero, and they will remain zero with harvesting. So there is no essential change in the community: there still is no reshuffle.

Suppose now that we indiscriminately harvest a dominance control-led community. In this case there are three qualitative possibilities for community structure, depending upon the values chosen for y and for Y/Y_{MS}, based on the following two distinctions:

1. Recall that our communities are random assemblages of species, with the competition coefficients α_{ij} ($i \neq j$) chosen at random from a positive real interval (e, f). Under indiscriminate harvesting, the upper bounds B_{is} on the competition coefficients which allow potential invasion (equation (10)) all have the same value, say B. From (9) we see that in case $B \leqslant e$ there will be no interpatch invasions at all. Using (10), (11), and the version of (12) which applies to harvesting structures S2, S3, and S4, we can write the condition $B \leqslant e$ as

$$\frac{1-(Y/Y_{MS})/4y}{(1+\sqrt{1-Y/Y_{MS}})/2} \leqslant e$$

or

(IV.14) $\quad y \leqslant \dfrac{Y/Y_{MS}}{4(1-e(1+\sqrt{1-Y/Y_{MS}})/2)}$

When this inequality is satisfied, there will be no interpatch inva-

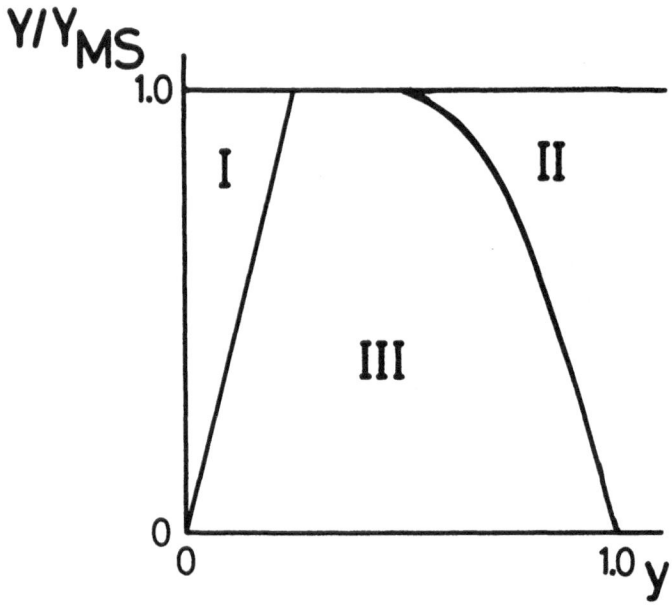

Figure IV.5. The qualitative structure of harvested communities assembled by random colonization from a pool of species for which e=0.

sions, hence no reshuffle: the harvested system will be in effect
founder controlled.

2. We have already remarked, through equation (13), that B will
never exceed 1 in value. However, it can attain the value 1, and when
it does the invasion probabilities will be exactly the same as they
are without harvesting, which means that community structure will be
substantially the same as without harvesting. This coincides with the
transition between harvesting structures S1 and S2 (figures 1 and 2):
when B=1 we have harvesting structure S1 (food-limitation of the harvesters). From equations (10), (11), and (12), this occurs when

(IV.15) $4y(1-y) < Y/Y_{MS}$.

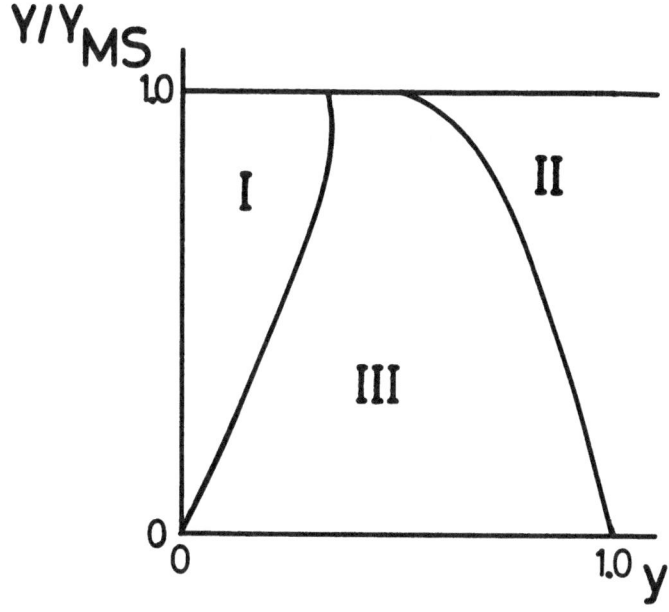

Figure IV.6. The qualitative structure of harvested communities assembled by random colonization from a pool of species for which e=0.5.

Thus, for each harvesting policy —— specified by giving a pair
$(y, Y/Y_{MS})$ of numbers, with $y \in (0, \infty)$ and $Y/Y_{MS} \in (0,1)$ —— the harvested
community will have one of three qualitative structures:

I. Founder controlled

II. Dominance controlled, food-limitation of harvesters

III. Dominance controlled, nonfood-limitation of harvesters.

When the inequality (14) is satisfied, we have case I. When the inequality (15) is satisfied, we have case II. Otherwise, we have case
III.

We can portray the occurrence of these three qualitative possibilities in a two-dimensional space with axes y and Y/Y_{MS}, as in figures
5, 6, and 7, where this is done for three different values of e. From
such a diagram one can get a rough idea of the qualitative effects of

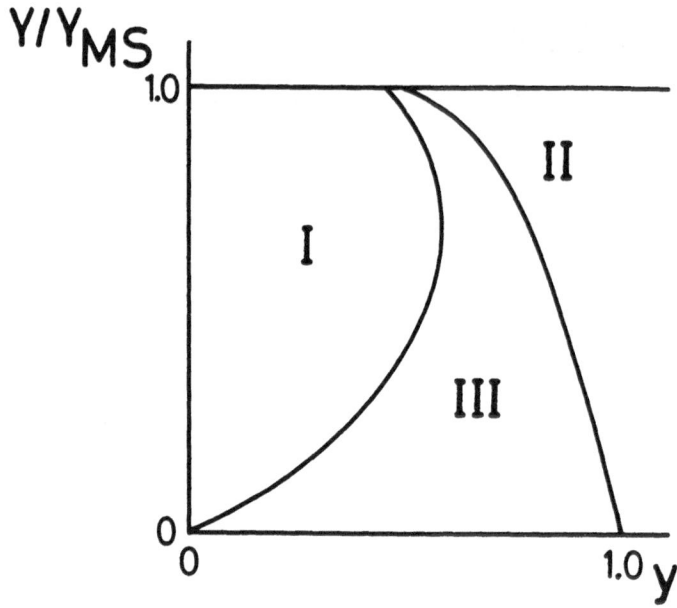

Figure IV.7. The qualitative structure of harvested communities assembled by random colonization from a pool of species for which e=0.9.

any change in harvesting policy.

   For example, suppose e=0.5 and that we vary the yield while hold-
ing the saturation density y constant. As we vary the yield we move
along a vertical line in figure 6. If, for example, we take y=0.25,
then for Y/Y_{MS}<0.6 the system is dominance controlled with nonfood
limitation of the harvesters, but when Y/Y_{MS} exceeds 0.6 the system
becomes founder controlled.

                                 *

   Figures 8-12 summarize the results of Monte Carlo simulations,
like those described in Sections III.3 and III.4, of the harvested
reshuffle process for communities formed by uniform colonization
(Section I.5). Each data point is an average over 100 randomly assem-

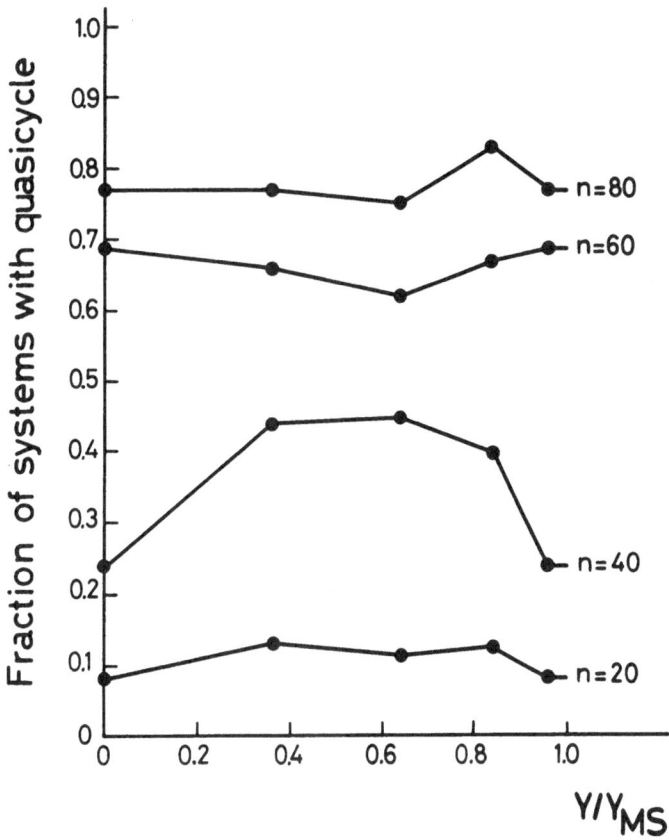

Figure IV.8. The fraction of harvested systems
with quasicycles as a function of the yield,
with the saturation density fixed at y=0.6.

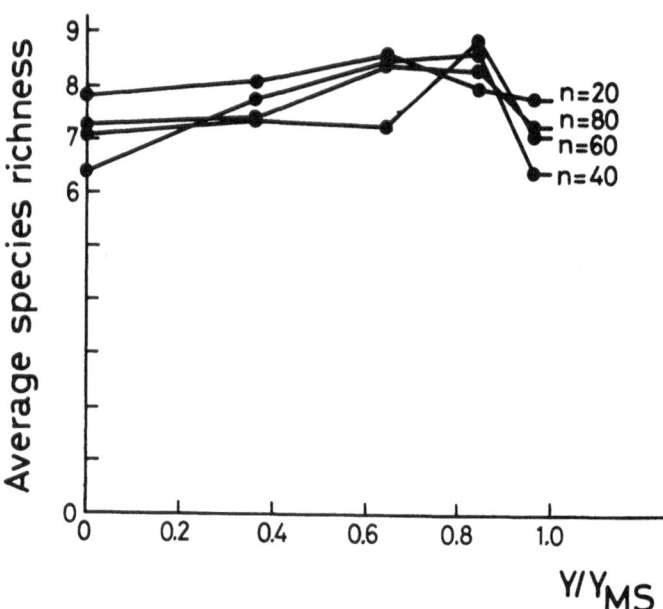

Figure IV.9. The average equilibrium species richness of harvested systems as a function of the yield, with the saturation density fixed at y=0.6.

bled systems, with each α_{ij} ($i \neq j$) chosen at random, with uniform probability distribution, from the interval (0.5,9.0), and each C_i from the interval (1,2).

*

In figures 8 and 9 the saturation density is held fixed at y=0.6. We see that with this value of y there is very little effect: the invasion probabilities never differ significantly from their values in the absence of harvesting.

*

On the other hand, when y is held fixed at 0.25 harvesting has dramatic effects, which can be seen in figures 10, 11, and 12. In this case there is a transition to founder control when Y/Y_{MS} exceeds 0.6, as can be seen also from figure 6.

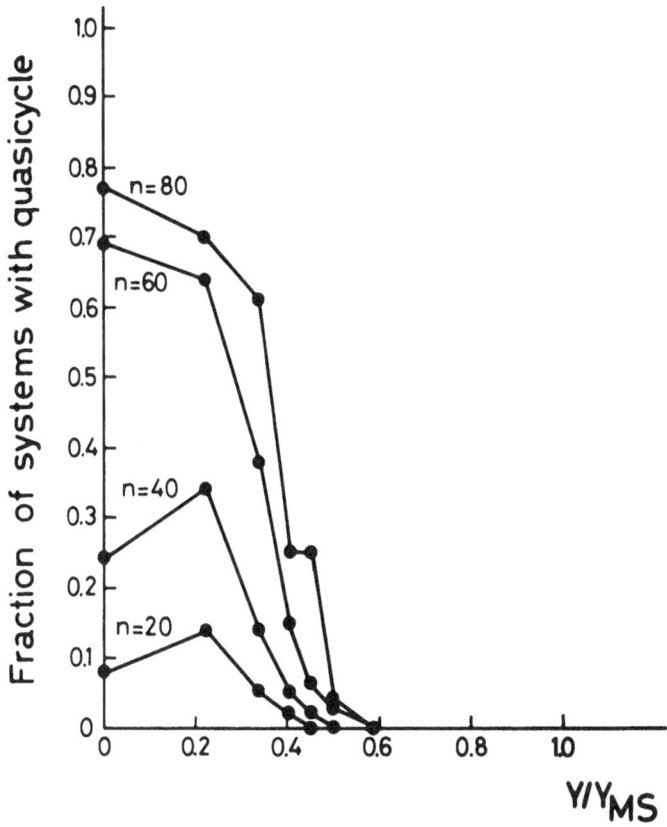

Figure IV.10. The fraction of harvested systems
with quasicycles as a function of the yield,
with the saturation density fixed at y=0.25.

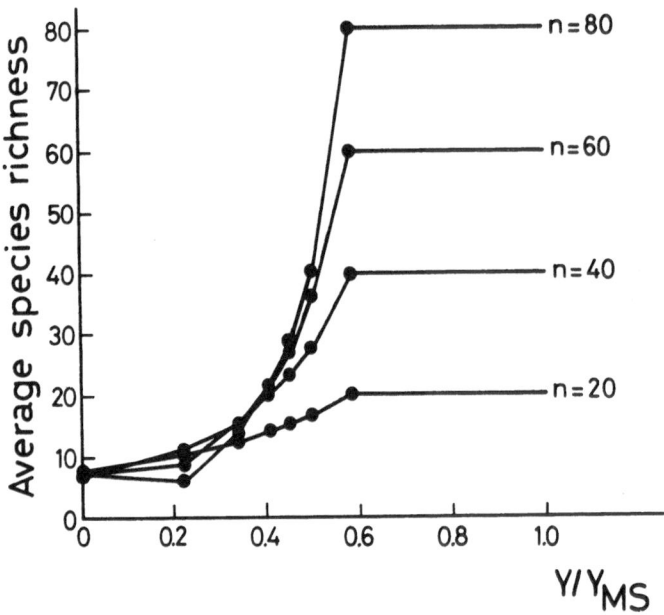

Figure IV.11. The average equilibrium species richness of harvested systems as a function of the yield, with the saturation density fixed at y=0.25.

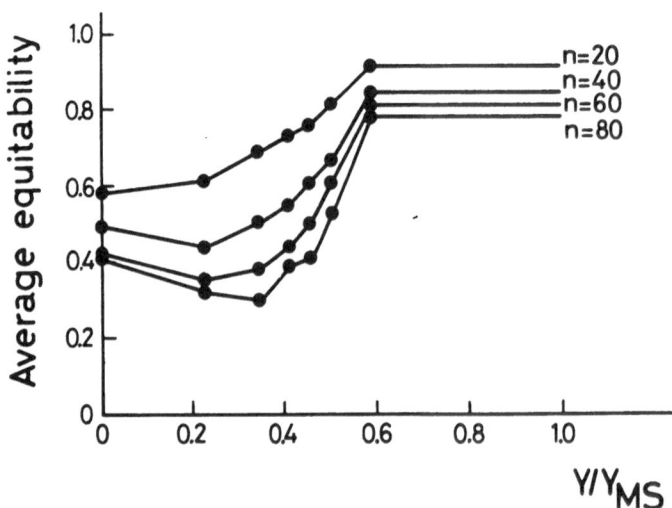

Figure IV.12. The average Simpson-Horn evenness of harvested systems at equilibrium as a function of the yield, with the saturation density fixed at y=0.25.

Summarizing: indiscriminate harvesting of dominance controlled systems tends to cause a decrease in the probability for quasicycles to occur, an increase in species richness, and an increase in evenness. These effects may be so small as to be unnoticeable, or they may be so large as to be conspicuous, depending upon the functional response of the harvesters (that is, in our present models, upon the values of the parameters y_i and Y/Y_{MS}).

IV.3.2. Selective harvesting of dominant species.

When there is a strong functional (hence numerical) dominant, it is advantageous for a grazer or predator to specialize in consuming the dominant species, and this frequently occurs in nature. We briefly consider here the consequences of this particular form of harvesting for community structure.

Suppose, then, that species d dominates several other species and that species d, and only species d, is harvested. Referring to equation (10), we see that all upper bounds B_{ij} will be unaffected by this harvesting, except for the bounds

$$B_{id} = \frac{1}{\kappa_d} \quad , \quad B_{ds} = 1-\zeta_d,$$

where i, s = 1,...,n. Thus all invasion probabilities (9) are unaltered, except for the probabilities p_{id} and p_{ds}.

The probability p_{id} for another species i to potentially invade a patch predominated by species d will increase under harvesting, and the probability p_{ds} for species d to potentially invade a patch predominated by some other species s will decrease under harvesting. The overall effect of these changes is to lessen the dominance of species d. Specifically:

1. If the system has a quasicycle in the absence of harvesting, this will be unaffected by harvesting. If it does not have a quasicycle in the absence of harvesting, it may have a quasicycle (involving species d) when it is harvested.

2. When species d is harvested it will in general be able to eliminate fewer other species from the system by dominating them; species

richness will tend to increase under harvesting.

3. Evenness will increase under harvesting whether richness does or not.

4. The Shannon-Weaver diversity index H (May 1975; Pielou 1975) will increase under harvesting whether richness does or not.

IV.3.3. Selective harvesting of nondominant species.

Sometimes a dominant species can manage to make itself unpalatable to predators; indeed, a species may be dominant precisely <u>because</u> it is unpalatable.

Assume, then, that some species d is not harvested, but all other species are. Then from (10)

$$B_{id} = 1-\zeta_i \quad , \quad B_{ds} = \frac{1}{\kappa_s},$$

where i, s = 1,...,n. This means that the invasion probabilities p_{id} are smaller, and the invasion probabilities p_{ds} larger, under harvesting. Therefore with harvesting species d may be able to dominate and eliminate from the system a number of species which it cannot dominate in the absence of harvesting. Species richness will, then, tend to decrease under harvesting.

Trends in other quantities will depend on the details of the harvesting.

IV.4. Species richness: observational.

There is a wealth of observational data in the natural historical literature concerning effects of harvesting on the species richness of various communities —— almost all of them sessile, and therefore involving spatial competition, presumably for reasons of observational convenience. We shall in this section compare the theory just developed with these observations.

First we summarize the theoretical results, then we discuss several pitfalls in comparing the theory with observations, and finally we look at the data, taking these pitfalls into account. We aim here for

comprehensiveness: this is the most thorough compendium of such observations of which the author is aware.

*

Since all of the field data are from dominance controlled communities, we review our theoretical results for that case only. It is
possible that harvesting could sometimes affect species richness
through its effect on quasicycles, but we will ignore this possibility
here, and summarize only the effects of harvesting on equilibrium communities.

Indiscriminate harvesting and selective harvesting of a dominant
species both tend to cause species richness to increase, and they
never cause it to decrease. Selective harvesting of nondominant species tends to cause species richness to decrease.

These statements assume that all yields are less than the maximum
sustainable yields. Of course, when maximum sustainable yields are
exceeded some species will be driven to (local) extinction and species
richness may very well decrease.

*

We list here some possible sources of confusion in comparing these
(or other such) results with observational data.

1. The most direct way to demonstrate these effects is to add harvesters to an unharvested system, or to remove harvesters from a harvested system, observing species richness before and after this change.
But the predicted effects will not necessarily be seen in "both direc-
tions". An example will make it clear what is meant here.

Suppose some dominant species is being selectively harvested from
one of these systems. If we remove the predators, the dominant species
can function unhindered and will, in general, eliminate one or more
other species from the system (Section 3.2); species richness will decrease after a time of the order of that characterizing the system dynamics.

Suppose that after this has occurred we replace the predators.
Then it is possible for the species which were eliminated in our ori-

ginal experiment once again to coexist with the dominant, if their propagules are available to reinvade the community. Therefore species richness will after a while increase, if 1) there is a sufficient influx of invaders from other habitat areas, and 2) we wait long enough for the (sometimes very slow) process of invasion and succession to take place.

We can summarize this by saying (with Watt 1957) that harvesting effects sometimes are dependent on the biotic potential (meaning the supply of invaders from other areas) of the area under consideration.

2. A major source of difficulty in understanding the observations is selectivity in harvesting. Most of the field studies involve grazing or predation, and even the most generalist grazers and predators have their dietary preferences. Very often dominant species are preferred, but seldom to the complete exclusion of all other species, as we assumed in Section 3.2.

We have in Sections 3.1, 3.2, and 3.3 discussed three idealized forms of diet, which were chosen so that most diets observed in nature are approximated by one of these three forms. But it is not always clear which idealization is the most appropriate; sometimes none of them may be a very good approximation, and calculations which are tailored to the habits of a definite consumer may need to be done.

Moreover, selective harvesting need not come about from fixed dietary preferences. "Switching" predators concentrate their efforts on the most abundant species at a given time (generally with a time lag). This saves foraging time, and indeed one can argue that switching is an optimal strategy in terms of energy budgeting (Emlen 1966, 1968). One might therefore expect switching to be widespread in nature, but most of the predators that will concern us do not seem to switch.

In the case of prey species which compete for renewable resources rather than for space, predation with switching can lead to an increase in species richness (Roughgarden and Feldman 1975). While we have not done any calculations for this case, it seems quite likely that predation with switching would also tend to increase species richness of communities where competition for space is important.

3. Another pitfall is obvious, but seems nevertheless to have caused considerable confusion in the literature on harvesting and

"diversity". The term "diversity" is used in the literature to mean
a number of different quantities, including for instance both species
richness and the Shannon-Weaver index H (May 1975; Pielou 1975). But
harvesting can affect these different measures of diversity quite dif-
ferently. For instance, we have already mentioned in Section 3.2 that
selective harvesting of a dominant species will cause H to decrease
even if it has no effect on richness.

Therefore, in studies of the effect of harvesting on "diversity",
one must be very careful about just how the word "diversity" is being
used. We discuss here effects of harvesting on species richness only.

4. Yet another obvious but very important pitfall is overharvesting: as already pointed out above, at very high harvesting rates some
species will be harvested to extinction, causing a decrease in species
richness, no matter what the character of the harvesting or of the
harvested community. This effect, which is from the theoretical point
of view quite trivial, should not be confused with other harvesting
effects which involve interspecific interactions and may, at lower harvesting rates, give different results.

5. Finally, it should be borne in mind that grazing animals, or
the manipulations of an experimenter, may have effects on the community
other than the primary effects of harvesting. For instance, animal
footprints may provide a source of environmental heterogeneity and
thereby increased species richness (such effects are discussed at
length by Harper 1977, Chapters 14 and 23); fences erected by an observer may interfere with dispersal of both predator and prey. We will
assume in our discussion that such effects have been taken into account
by each observer.

*

Let us, then, proceed to the field data. We began this chapter
by citing Darwin's (1859) lawn-mowing experiment. This would seem to
be a clear example of an increase in richness due to indiscriminate
harvesting (for another point of view, see Van Valen 1974). Our other
examples involving terrestrial plant communities come from grazing and
are therefore complicated by the possibility of dietary selection.

Grazing by rabbits has long been known to have profound effects
on grasslands (this is nicely put in an (English) historical context

by Thomas (1960)). Tansley and Adamson (1925) erected rabbit-proof
exclosures in chalk grasslands near the Hampshire-Sussex border, and
found that after six years species richness in the exclosures had declined dramatically, with Festuca ovina achieving strong dominance.
Hope-Simpson (1940), visiting these same areas some fifteen years
later in order to study the effect of grazing by rabbits on succession,
pointed out that woody plants were unable to invade areas subject to
heavy grazing, so that the very identity of these areas as grassland
is due largely to grazing.

These authors also found that on steep northern or otherwise
shaded slopes grazing by rabbits permitted mosses (which the rabbits
seldom graze) to become dominant. A similar effect was observed by
Summerhayes (1941) when he excluded voles by means of cages from portions of Scottish grassland. The only clear change in species composition observed in his experiments was an almost complete elimination
of mosses from the caged-in (and therefore ungrazed) areas.

In another experiment involving rabbit-proof exclosures, this time
in grasslands in Breckland, Watt (1960) found that inside the exclosures species richness declined to about one-half its former value over
periods of 5 to 6 years, accompanied by the establishment of strong
dominants. Watt argues that the grazing was, in our terminology, indiscriminate in this case, so this can perhaps be taken as supporting
our results in Section 3.1.

It is instructive to compare these experiments with another by
Watt (1957), in which a fence separated grazed from ungrazed grassland.
In this case the ungrazed area was found to contain more species, in
seeming contradiction with the exclosure experiments and with our theoretical results. As pointed out by Watt, the difference here is one
of biotic potential: in this experiment there was time —— at least 30
years —— for a number of species particularly sensitive to rabbit
grazing to become established. These are species which would take very
long indeed to invade a rabbit-proof exclosure, for it is surrounded
by grazed area, in which these species are kept, at best, extremely
rare.

Harper's (1969) classic studies of grazing by sheep in Britain
show strong effects of dietary preference. He found that heavy grazing
by sheep increased the richness of pastures originally composed largely
of species palatable to the sheep, but decreased the richness of pas-

tures composed largely of unpalatable species, in agreement with our theoretical results.

The 90% reduction in species richness of lichen communities due to zinc pollution documented by Nash (1975) is almost certainly an overharvesting effect.

A rich body of data comes from intertidal zones. We have already mentioned at the beginning of this chapter Paine's (1966, 1974) experiments on removal of the predatory starfish _Pisaster ochraceus_ from intertidal communities. Dietary selection by _Pisaster_ does not fit very well into any of our three idealized categories, for while it does have rather catholic tastes, it shows some preference for mussels, including the competitive dominant _Mytilus californianus_ (Paine 1966; Landenberger 1968). Thus it in some sense straddles our two categories "indiscriminate" and "selective of dominant".

This is not a serious ambiguity for our present purposes, for the theory makes the same prediction in both of these categories, namely that species richness should increase with harvesting, and this is exactly what Paine observes. This effect of _Pisaster_ has been confirmed in observations by Dayton (1971, 1973a), and the same effect has been seen in similar experiments with predatory starfish in New Zealand (Paine 1971) and in Chile (R. T. Paine, private communication).

Connell (1970b), in observations of the intertidal zone of San Juan Island, Washington, USA, found that in situations with at least moderate water movement predators were common and fed heavily on both of the barnacle species _Balanus glandula_ and _B. cariosus_, enabling them to coexist. Menge (1976) showed by means of predator exclusion cages that predation by the snail _Thais lapillus_ makes it possible for _B. balanoides_ and the mussel _Mytilus edulis_ to coexist in the New England (USA) rocky intertidal. Harvesting is probably best thought of as indiscriminate in both of these experiments, and the results are again in agreement with theory.

The experiments of Lubchenco Menge (1978) on grazing of algae by the marine snail _Littorina littorea_ provide a marine analogy to Harper's observations on the effects of grazing by sheep, discussed above: when _Littorina_'s preferred food is competitively dominant, richness increases with grazing; when the preferred food is competitively inferior, richness decreases with increased grazing.

Nicotri (1977) found that grazing by intertidal gastropods on microalgal communities, both on natural rock surfaces and on artificial substrates, tended to cause a decrease in microalgal species richness. This result is difficult to relate to theory, because it is unclear for what (if anything) the algae compete; we mention it here because they may conceivably compete for space. The natural rock surfaces are certainly overharvested, and selective overharvesting could also be the explanation for the results from artificial substrates.

Both Paine and Vadas (1969) and Dayton (1975b) found an increase in macroalgal richness after removal of sea urchins from portions of intertidal zones; thus grazing had here caused species richness to diminish. However, as already remarked in Section 1, sea urchins tend to overgraze algae, and this seems the most likely explanation for these observations.

Porter (1972, 1974) and Glynn (1976) observed quite different effects due to predation on coral reefs in the eastern Pacific Ocean by the starfish Acanthaster planci: Porter found that predation caused an increase in species richness, Glynn a decrease. Now, it is known that the competitive dominant in these reefs, Pocillopora damicornis, is normally a preferred prey species for Acanthaster, on which basis our theory would lead one to expect an increase in richness from this predation, as observed by Porter. However, Glynn observed that Pocillopora corals in his reef were occupied by crustacean symbionts which often drove away Acanthaster which were attempting to feed. Thus in his system Pocillopora was selectively avoided. Under these circumstances the theory indicates that predation should cause species richness to decrease, which is the effect observed by Glynn.

Finally, Slobodkin (1964) found in laboratory experiments that two species of hydrida, when competing for light, could coexist only when harvested at quite high rates (periodic removal of 90% of both population increments since the previous census). This harvesting is clearly food-limited, and the result is in agreement with Slatkin's (1974) theoretical treatment of that case.

*

While interpretation of these observations in terms of theory is seldom entirely unambiguous, overall agreement with theory seems to be excellent. At the very least, there are no obvious conflicts between theory and observation.

IV.5. Disturbance.

Sessile species are especially vulnerable to physical disturbance, and this surely plays an important part in the structure of sessile communities. For instance, wave action and battering by floating logs has the effect of opening up space in intertidal communities for fresh colonization from the surrounding community or from elsewhere; forests suffer damage by fire and wind. This sort of disturbance differs from what we have called harvesting in that it removes biomass in a patchy (in space or in time or in both) fashion.

Empirically, the effects of disturbance on species richness are very similar to those of harvesting. We have already cited in Section III.5 suggestions that rain damage in tropical forests and fire in temperate forests help maintain species richness in these communities. In Pacific northwest (USA) intertidal communities, damage from waves and from floating logs has effects very similar to those of starfish predation (Dayton 1971, 1973a, 1975a), and wave damage also appears to maintain coral richness off the Hawaiian coast (Grigg and Maragos 1974).

Osman (1977) found that periodically turning over the rocks in a rocky subtidal zone in Massachusetts (USA) had clear effects on the structure of epifaunal communities on the rocks, with species richness increasing to a maximum and then decreasing as the rate of disturbance increased. Abele (1976) found that the number of decapod crustaceans supported by coral reef heads on the Pacific coast of Panama increases as environmental irregularity increases. Cairns, Dickson, and Yongue (1971) suspended artificial sponges in a lake and observed the protozoa which colonized them; they found that sponges which were squeezed periodically had significantly higher species richness.

Incorporating this sort of disturbance into our models would require knowledge of the time scale for reshuffles: if this time scale were short compared to the time scale for recurrent disturbance the effects of disturbance would be relatively weak; if the time scale for disturbance were shorter, reshuffles would become more or less irrelevant and the community would be structured primarily by accidents of disturbance and recolonization (see Levin and Paine 1974, 1975).

IV.6. Summary.

This chapter discusses the influence of harvesting (by which we
mean any more or less continuous (in space and in time) removal of
biomass, be it by grazing, by predation, or by intentional or unintentional human action) on the communities studied in Chapters II and III.

Section 1 establishes our mathematical framework for harvesting,
and distinguishes several different qualitative dynamical structures
for the harvesting of one population, with examples from the natural
historical literature.

In Section 2 the rules for reshuffles in harvested communities
are derived.

The influence of harvesting on the behavior of these models is
discussed in Section 3. Three forms of dietary selection are considered: indiscriminate harvesting (all species harvested at roughly comparable rates) in Section 3.1, selective harvesting of a competitive
dominant in Section 3.2, and selective harvesting of nondominant species in Section 3.3. The case of indiscriminate harvesting is treated
the most thoroughly, both because it is intuitively the least obvious
case and because it is important as a paradigm for general harvesting
pressure from higher trophic levels, possibly involving many harvesters.

It is found that indiscriminate harvesting and selective harvesting of a dominant tend to cause species richness and evenness to increase. Selective harvesting of nondominant species tends to cause
species richness to decrease; its effect on evenness depends on the
details of the harvesting. Indiscriminate harvesting causes a decrease, and selective harvesting of a dominant species an increase,
in the probability of quasicycles. The effect of selective harvesting
of nondominant species on this probability depends on the details of
the harvesting.

These effects all obtain at moderate harvesting rates. Different
effects can, of course, be obtained when harvesting rates are so high
that the harvested species can no longer maintain positive growth.
These "overharvesting" effects, while quite trivial theoretically, are

frequently observed.

Section 4 discusses the relation between harvesting and species richness in sessile communities from an observational point of view. Some pitfalls in comparing theory with observation are pointed out, and the observational literature is comprehensively reviewed. Agreement between theory and observation appears to be good.

Observations of patchy (in space or in time or in both) disturbance are reviewed in Section 5.

Chapter V. EPILOGUE

When competition for space is important and the possibilities for
niche differentiation are restricted as described in Section I.4, the
resultant communities fall into two broad categories (we discuss the
consequences of dropping these restrictions in Section 1). These can
conveniently be envisioned in terms of the subsequent development of
a community after the colonization of free space by a number of species, with at first one species per patch.

This development takes one of two forms: either the original colonizer in each patch maintains its predominance there (with other species present only in small numbers), or in some patches other species,
from other patches, invade and achieve predominance. Communities
formed in the first way we have called founder controlled (Chapter II);
those formed in the second way, dominance controlled. The terminology
here is meant to convey the main factors which govern community structure. This structure is itself quite different in the two cases.

The structure of a dominance controlled community is governed
primarily by functional dominance relations among the species. Because
of these relations, some colonizing species are eventually eliminated
from the community entirely, while others may achieve conspicuous abundance. Thus a rocky intertidal community in the Pacific northwest of
America will often consist of what appears to be a large bed of the
mussel Mytilus californianus interspersed here and there with patches
of other species (Paine 1974; Dayton 1973a), and abandoned farmland in
the American midwest will often develop into forests in which beeches
and maples are so common that such a forest is frequently referred to
as a "Beech-Maple Association", although many other species besides
these two can be seen there (Williams 1936).

The vast majority of sessile communities documented in the natural

historical literature exhibit this typical behavior of dominance control.

In each locality there is a pool of species which are potential
colonizers of any space that may be in one way or another cleared, and
there is a probability p that one of these species chosen at random
can invade a patch predominated by another randomly chosen species.
In Chapter III we found that the structure of communities formed by
random colonization from this pool of potential colonizers is determined in its essential features largely by the probability p. This is
probably the focal perception of the study reported in this volume.

Founder controlled communities (for which the probability p is
zero) lack this definite structure: their structure is governed mainly
by accidents of colonization. We will, in Section 3, discuss the species rich wet lowland tropical forest as a possible example.

Another important determinant of community structure is harvesting, by which we mean some regular removal of biomass from a community,
be it by grazing, by predation, or by intentional or unintentional
human action (Chapter IV). Especially for dominance controlled communities, harvesting can have profound effects, including, usually,
increased species richness and evenness. These changes come about
through a reduction, due to harvesting, in the effective value of the
probability p. In extreme cases p can be reduced to zero, so that the
community is in effect founder controlled.

*

Chapters II-IV develop the themes just sketched in considerable
detail and compare specific theoretical results with observations when-
ever possible. In the epilogue which now follows some broad theoretical implications of these ideas are considered. Because of its necessarily tentative character, the discussion here is sometimes more spe-
culative than that in earlier chapters.

V.1. Spatial competition and the niche reconsidered.

We saw in Section I.2 that niche relations —— the partitioning
of resources —— play an important role in communities of sessile organisms, especially terrestrial plants. On the other hand, we found

in Chapter III that even when the possibilities for niche differentiation are severely restricted, spatial competition, through dominance
relations, is capable of producing much of the observed structure of
these communities.

Resource competition and spatial competition occur in subtle combination in sessile (and other: see below) communities, and a full
understanding of these communities must take into account both kinds
of competition. The reaction-dispersal models of Section I.3 combine
these processes quite naturally; indeed, in order to pursue our goal
of studying more or less pure spatial competition we had, in Section
I.4, to explicitly restrict the possibilities for niche differentiation
in these models. Dropping those restrictions produces models which
nontrivially combine niche and dominance relations.

As the probability p for a competition coefficient to be less than
1 increases, resource competition becomes progressively more important
relative to spatial competition. In the extreme case where all alphas
are less than 1 spatial competition may be only of minor importance.
For instance, Casten and Case (1978) found that local stability of the
multi-species equilibria in systems with pure exploitation competition
is unaffected by diffusion, which implies that for these systems there
is probably no spatial patterning.

One can get some idea of the relative importance of resource competition and spatial competition in a community by looking at the influence of harvesting on it, for harvesting can have quite different
effects on the two kinds of competitive relations. Indiscriminate
harvesting with nonfood-limitation of the harvesters is very well
suited to this purpose, for this kind of harvesting gives opposite
effects: when the harvested species compete only by partitioning resources, this harvesting causes species richness to decrease (Van Valen
1974; Yodzis 1977); but when the harvested species compete primarily
for space, this harvesting causes species richness to increase (Section
IV.3.1).

Data from a variety of sessile communities, reviewed in Section
IV.4, suggest on this basis that spatial competition tends to be the
more important factor in determining species richness in these communities. It would be most interesting to contrast this with corresponding data from communities in which spatial competition is clearly of
no importance; unfortunately, very little such data can be found in the

literature.

There is quite a lot of data concerning effects of predation by
fish on zooplankton diversity (for instance, Brooks and Dodson 1965;
Hall, Cooper, and Werner 1970), but this predation is highly selective
(Brooks and Dodson 1965; Eggers 1977) and its effects are easily understood in terms of this selectivity. The only appropriate data known
to the author are found in the study by Adicott (1974) of predation by
mosquito larvae on protozoan communities in pitcher plant leaves.
Competition for space is clearly unimportant here, and Adicott indicates that the harvesting is, in our terms, indiscriminate. He finds
a significant negative correlation between predation intensity and
species richness, in agreement with the conclusions of Van Valen (1974)
and Yodzis (1977), and in contrast with the results from sessile communities. (Of course, the possibility remains open that this is simply an overharvesting effect. Only further observations can settle
the issue.)

*

Anomalies in the conventional picture of niche relations are sometimes due to spatial effects. Thus Platt and Weis (1977) found that
species packing in a guild of fugitive prairie plants was closer than
predicted by theory, until spatial dispersion was explicitly taken into
account. They managned to express spatial relations in their system
as an additional niche dimension, but it is unlikely that all spatial
effects can be treated in this way.

A similar anomaly involves coral reef fish: for some guilds of
these fish there appears to be very extensive overlap in any niche
space based on utilization of food resources. Roughgarden (1974)
points out that this can be explained if the resource utilization
functions are thick-tailed, but this is a very difficult hypothesis
to test. However, Roughgarden (private communication; to be published)
has in the meanwhile assembled other data which support a niche interpretation. On the other hand, Sale (1977) has recently offered an
interpretation which does not involve competition at all.

Another viewpoint, in some sense intermediate between these two,
would emphasize that many coral reef fish display interspecific territoriality, and that a good case can be made for considering these fish

to be space-limited (Sale 1972; Smith and Tyler 1972). Indeed, close examination of the spatio-temporal distribution of these fish in an Australian reef habitat by Sale (1975) suggests a dominance controlled system (see Section III.3), with coexistence facilitated by territoriality.

V.2. Species packing and the relation between plant and animal diversity.

One of the major goals of theoretical ecology is to identify and perhaps even to quantify those factors which determine how many species can be "packed" into a given habitat. So far as sessile communities are concerned, we have seen (particularly in figure III.5) that competition for space, through dominance relations, is an important one of these factors. This would suggest, for example, that plant species diversity is determined at least in part by competition for space. But this may have implications also for other aspects of ecosystem diversity.

In a classic paper, Hutchinson (1959) argued that animal species diversity is in part a function of plant species diversity. This notion is of obvious interest to the student of spatial competition; let us, then, examine the relevant field data.

Probably the best known investigation is that of MacArthur and MacArthur (1961), which reached a somewhat negative conclusion. These authors found that, while insectivorous bird species diversity in North American forests does correlate well with plant species diversity, it correlates better with foliage height diversity, a measure of floristic _structural_ diversity. (Of course, there is also a correlation between structural and species diversity.) A number of subsequent studies (MacArthur, Recher, and Cody 1966; Recher 1969; Cody 1970; Karr 1971; Karr and Roth 1971; Pianka and Huey 1971; Blondel, Ferry, and Frochot 1973) confirmed the correlation between floristic structural diversity and bird species diversity.

However, these studies did not measure plant species diversity and therefore do not impinge upon the issue of whether structural or species diversity of plants is a more important determinant of bird species diversity. A recent study by Abbott, Abbott, and Grant (1977) treats both structural and species diversity of flora in the Galapagos

Islands, and finds in a multiple regression analysis that plant species diversity is more important than structural diversity in accounting for the diversity of the total land bird fauna.

A major difference between the study of MacArthur and MacArthur and that of Abbott, Abbott, and Grant is that while the former study involved insectivores, the bird species dealt with in the latter study eat seeds and fruits. These studies are, then, consistent with the hypothesis that plant species diversity is a major determinant of herbivore species diversity, but is a less important influence on species diversity at higher trophic levels.

The case of herbivorous insects is somewhat clearer. Since plants (being unable to escape their enemies) commonly produce a wide variety of toxins, it is advantageous for small herbivores to specialize in the detoxification of only a few of these substances and therefore to consume only a few species of plants. It comes, then, as no surprise that insect herbivores tend to be specialized (Brues 1946; Whittaker and Feeny 1971). It is therefore extremely plausible that greater plant species diversity will allow greater insect species diversity.

But the causal chain is not so simple: as insects evolve the ability to detoxify existing substances produced by plants, so must plants evolve new toxins in order to protect themselves (e.g., Whittaker 1972; Green and Palmbald 1975). Plant species diversity and herbivorous insect species diversity are linked by a kind of evolutionary feedback. Thus, Murdoch, Evans, and Peterson (1972) found a strong correlation between plant and insect species diversity (both richness and evenness) in three Michigan old fields. (However, they also found strong correlations of both of these quantities with floristic structural diversity.)

On the other hand Rosenzweig and Winakur (1966) found no correlation between species diversity of (mostly granivorous) desert rodents and plant species diversity, though they did find some correlation with plant structural diversity. Finally, moving up the trophic ladder again, Pianka (1967) found no correlation between plant species diversity and lizard species diversity in various desert communities, all but one of the lizard species treated here being carnivores.

*

It seems reasonable to say that plant species diversity is often an important influence on herbivore species diversity, but is a less important influence at higher trophic levels.

The implications of this for total ecosystem species diversity are difficult to assess. Part of the trophic level conception of ecosystems (Elton 1927; Lindeman 1942) is the generality that there tends to be greater species diversity at lower trophic levels, which would suggest that those factors which determine diversity at the lower trophic levels are the most important factors for total diversity. However, it is now clear that the whole trophic level framework is an oversimplification which needs seriously to be reassessed (for instance, Petrusewicz and Macfadyen 1970; May 1976).

This does not rule out the possibility that parts of this framework may survive reappraisal. For instance, the categories primary producer, herbivore, omnivore, and carnivore are likely to remain useful. If there is, after all, a concentration of diversity in the first three of these categories, then competition for space may play a major role in determining overall ecosystem diversity.

V.3. Evolutionary considerations and the tropical rain forest.

We have seen repeatedly in this study that the probability p (equation (III.4)) which characterizes the pool of available colonizing species in a given locale is a major determinant of structure in sessile communities which are assembled from those colonizers. The quantity p is, in turn, determined by the course of evolution in each locale.

As the species pool in some locale evolves by natural selection, so will the interactions among these species —— in particular, the probability p —— evolve. We are still far from a complete understanding of such coevolutionary processes (for a lucid review of the current state of the art, see Levin 1978b), but we can get some idea of what to expect as follows.

The probability p is an inverse measure of "competitive strength": the smaller the value of p for some pool of colonizers, the larger is the probability that a randomly chosen species from this pool will hinder the growth of a competitor randomly chosen from the pool more than it hinders its own growth. Thus the value of p will be determined

by tradeoffs between competitive ability and other beneficial attributes which, however, like competitive ability, cost time and energy.

The balance finally arrived at by each species among its various
energy expenditures will depend partly on environmental factors such
as climate and the availability of resources. Some localities are,
in some sense which remains to be precisely formulated, more "favorable" than others. A more favorable physical environment should allow
a greater expenditure of energy on competition, as argued, for instance, by Dobzhansky (1950) and Williams (1964). This suggests that
p values should tend to be smaller is more favorable localities.

*

The wet tropical lowlands would seem to provide a singularly
favorable environment for plants: rainfall and sunlight, both essential requirements for plant life, are plentiful, and the climate is
significantly less variable than that in temperate zones. Thus, p
values should tend to be lower in the tropics than in other regions,
from which it follows (Sections III.4.2, 4.3) that plant diversity
should be higher in the tropics.

In fact, a conspicuous increase in diversity as one goes from
the poles to the tropics is one of the major trends of biogeography
(for instance, Pianka 1966). The reasoning that we have just sketched
amounts to at least a partial explanation for this trend; let us call
our explanation the "dominance hypothesis" for the latitudinal diversity gradient.

This hypothesis is quite distinct from the "competition hypothesis" (Pianka 1966) of Dobzhansky (1950) and Williams (1964), even
though these authors develop their ideas from the same starting point
that we have used. They argue, as we do, that for climatic reasons
competition should be "keener" in the tropics, but they apply this
observation to resource competition, suggesting that keener competition should result in narrower niches and thus in greater diversity.

It is interesting to consider the relation of these two hypotheses
to yet a third: the "predation hypothesis" of Gillet (1962), Paine
(1966), Janzen (1970, 1972), and Connell (1970b). These authors argue
that (again for climatic reasons) the general level of predation, especially on vegetation, is higher in the tropics than in other regions,

and they suggest that the effect of this is to alleviate competition
and allow more species to coexist.

Pianka (1966) remarks that the competition hypothesis and the
predation hypothesis function in quite opposite ways, and this point
is borne out by the work of Yodzis (1977) on indiscriminate harvesting of resource-limited communities: the processes envisioned in these
two hypotheses, if both at work, would tend to cancell one another out.

However, we have seen (Section IV.3.1) that in the context of competition for space indiscriminate harvesting (which can be thought of
as a rough approximation to general predation pressure from many specialized predators, as in the tropics) does indeed cause diversity to
increase, supporting the predation hypothesis at least where competition for space is concerned. Moreover, this comes about because harvesting causes the effective value of p to decrease. Thus the dominance and predation hypotheses complement one another: the processes
invoked in both of these hypotheses affect community structure by
causing the effective value of p to decrease.

*

We suggest, then, that p should tend to have smaller values in
tropical rain forests than in forests elsewhere. It could even be
that p is zero or nearly zero in these forests so that they are founder
controlled or something quite like it. Let us examine the evidence.

Firstly, some forests in the tropical lowlands are very similar
in community structure to temperate zone forests, with, for example,
dominance by one or two species. However, these forests occur on poor
soils or in habitats which are otherwise marginal, and it is generally
believed (e.g., Richards 1952) that the character of these forests is
very strongly influenced by the necessity to evolve tolerance for specific unfavorable conditions. (For another viewpoint, see Connell
1978).

The forests which are generally thought of as typical tropical
rain forests occur on well-drained soils, and it is to these forests
that our dominance hypothesis should apply (recall that the hypothesis
requires a "favorable" environment). These differ in their community
structure from temperate zone forests in the following ways (Eggeling
1947; Black, Dobzhansky, and Pavan 1950; Richards 1952; Pires, Dobzhan-

sky, and Black 1953; Cain, de Oliviera Castro, Murca Pires, and da Silve 1956; Poore 1964, 1968; Ashton 1969; Paijmans 1970; Knight 1975):

1) Species richness is typically an order of magnitude greater in the tropical rain forest than in temperate forests;

2) Dominance is conspicuously absent in the tropical rain forest, with Simpson-Horn evenness (equation (III.11)) typically two orders of magnitude greater than in temperate forests;

3) Species-area curves show little sign of levelling off at large areas in the tropical rain forest.

These are precisely the characteristics that one expects of a founder controlled community (p zero), or of a weakly dominance con-trolled community (p very small but not zero). Many observers (for instance, Aubréville 1938; Richards 1952; Hewetson 1956; Schulz 1960; Poore 1964, 1968; Rollet 1969; Fox 1970; Webb, Tracey, and Williams 1972) have remarked that reoccupancy of openings produced by treefalls has in these forests a large random component, reinforcing the notion of small p values.

Spatial competition theory offers, then, a simple and biologically sensible explanation not only for the remarkably similar structure of so many sessile communities, but also for the unique character of the tropical rain forest.

Appendix A. PERTURBATION THEORY

In this appendix, we first discuss the following mathematical problem: given a system of N first-order ordinary differential equations, which depend on some parameter q,

(A.1) $\dfrac{dx_i}{dt} = F_i(x;q)$, $i=1,\ldots,n$,

how do equilibria of the system vary as we vary the parameter q.   In Section 1 we discuss the behavior of the equilibrium values of the x's. In Section 2 we discuss the behavior of the eigenvalues of the linearized stability matrix; in particular we obtain here formulae for the lowest order perturbations to semisimple eigenvalues of arbitrary complex matrices which are analytic in the perturbation parameter. Then, in Section 3, we use these methods to study local stability of some equilibria of our reaction-dispersal systems.

A.1. Perturbed equilibria.

Consider a system of the form (1). In this equation (as elsewhere) x stands for the N real variables $x_1,\ldots,x_N$, where N is a positive integer. The domain of definition of the real parameter q can without loss of generality be taken to include the value 0. We shall refer to the system (1) with q=0 as the <u>unperturbed system</u>, and shall regard the act of letting q become nonzero as a <u>perturbation</u> of this system.

Let $\hat{x}$ be an equilibrium point of the unperturbed system:

$$F_i(\hat{x};q)=0 \quad , \quad i=1,\ldots,N.$$

We want to know what happens to this equilibrium as we let q take on nonzero values. In general, we expect that as we vary q in a neighborhood of 0 the equilibrium values of the x's will move around in a neighborhood of $\hat{x}$: for each value of q in a neighborhood of 0 we expect a <u>perturbed equilibrium</u> x(q), which must be a real solution of the N equations

$$(A.2) \qquad F_i(x(q),q)=0 \quad , \quad i=1,\ldots,N,$$

with $x(0)=\hat{x}$.

However, it is not necessarily true that a perturbed equilibrium will exist for each value of q, nor even for values of q which are arbitrarily close to 0. Even if an equilibrium exists, it may not be unique.

Consider for example the two-dimensional system

$$\frac{dx_1}{dt} = x_2 - x_1^2 - q$$

$$\frac{dx_2}{dt} = x_2 + x_1^2 + q.$$

This has the unperturbed equilibrium $(\hat{x}_1,\hat{x}_2)=(0,0)$ when q=0. In this case we can explicitly solve the equations (2) to obtain

$$x_1(q)=\pm\sqrt{-q} \quad , \quad x_2(q)=0.$$

Thus if q takes on (even arbitrarily small) positive values, there is no equilibrium (recall that the variables x_1, x_2 are real), and if q takes on negative values, there is not one equilibrium, but two.

Usually it will not be possible to explicitly solve the equations (2), so we need conditions which can be routinely checked to tell us whether such peculiarities will occur. These conditions are easily found:

Proposition A.1. Suppose 1) the N functions $F_i(x; q)$, i=1,...,N, are analytic in all N+1 variables $x_1,\ldots,x_N,q$. in some neighborhood of $(x,q)=(\hat{x},0)$, 2) the matrix with elements

(A.3) $$A_{ij}^{(0)} = \frac{\partial F_i}{\partial x_j}(\hat{x},0)$$

is nonsingular. Then for q in a sufficiently small neighborhood of 0, there exists a unique family of perturbed equilibria x(q) with $x(0)=\hat{x}$. Moreover, the N functions $x_1(q),\ldots,x_N(q)$ are analytic for q in a neighborhood of 0.

Proof. This is precisely the analytic version of the Implicit Function Theorem of classical analysis. See, for instance, Bochner and Martin (1948) or Dieudonné (1969). ∎

If the hypotheses of this proposition are satisfied, then x(q) can be expanded as a power series in q in a neighborhood of q=0. Writing

(A.4) $$F_i(x;q)=F_i^{(0)}(x)+F_i^{(1)}(x)q+ \cdots$$

we find by direct substitution

(A.5) $$x_i(q)=\hat{x}_i-\Sigma_j A_{ij}^{(0)-1}F_j^{(1)}(\hat{x})q+ \cdots$$

where $A_{ij}^{(0)-1}$ are the elements of the inverse matrix to the matrix defined by equation (3):

$$\Sigma_k A_{ik}^{(0)}A_{kj}^{(0)-1} = \Sigma_k A_{ik}^{(0)-1}A_{kj}^{(0)} = \delta_{ij},$$

where δ_{ij} is the <u>Kronecker delta</u>

(A.6) $\delta_{ij} = \begin{cases} 1, & i=j \\ 0, & i \neq j. \end{cases}$

Existence of this inverse matrix is guaranteed by our assumption in Proposition 1 that $A^{(0)}$ is nonsingular.

A.2. <u>Perturbed stability matrices</u>.

We consider next the <u>perturbed stability matrix</u> $A(q)$, with elements

(A.7) $A_{ij}(q) = \dfrac{\partial F_i}{\partial x_j}(x(q);q).$

Notice that

$$A_{ij}(0) = A_{ij}^{(0)},$$

where the <u>unperturbed stability matrix</u> $A^{(0)}$ has the elements (3).

The matrix $A(q)$ contains considerable information about local stability properties of the perturbed equilibrium. For instance, the linearized stability condition — that all eigenvalues of $A(q)$ have negative real parts — is a sufficient condition for asymptotic stability in the sense of Liapunov (LaSalle and Lefschetz 1961; Hahn 1967). In addition, since its eigenvalues enter the solutions of the linearized approximation to equation (1) multiplying the time in exponentials (Coddington and Levinson 1955) it gives us, when the equilibrium is linearization stable, an estimate of how rapidly equilibrium is approached.

<u>Corollary A.1</u>. Under the assumptions of Proposition 1, the elements $A_{ij}(q)$ of the perturbed stability matrix are analytic functions of q, for q in some neighborhood of 0.

Proof. Each partial derivative of an analytic function is analytic; an analytic function of analytic functions is analytic (Bochner and Martin 1948; Dieudonné 1969). ▌

Therefore, we can for q in a neighborhood of 0 write A(q) in the form

$$(A.8) \qquad A_{ij}(q) = A_{ij}^{(0)} + A_{ij}^{(1)} \times q + \cdots$$

where, one easily finds,

$$A_{ij}^{(1)} = - \sum_{k,m} \frac{\partial^2 F_i^{(0)}}{\partial x_j \partial x_k}(\hat{x}) A_{km}^{(0)-1} F_m^{(1)}(\hat{x}) + \frac{\partial F_i^{(1)}}{\partial x_j}(\hat{x}).$$

*

The stability information contained in A(q) can be expressed in terms of its eigenvalues. In general, it is not however possible to solve for these eigenvalues explicitly; thus one would like to have a power series expansion for the eigenvalues and a method for finding at least the lower order coefficients in this expansion. This is the content of perturbation theory for matrices, which we shall discuss in the remainder of this section.

Some readers may have been annoyed by our careful proofs that everything of interest to us, at least up to now, is analytic. These readers may now be inclined to say: we know that A(q) is analytic in q; just assume that the eigenvalues are also analytic, write formal power series for them, put these into the eigenvalue equations, and solve for the coefficients. But such an approach is, in general, doomed to failure. Consider, for example, the matrix

$$(A.10) \qquad \begin{bmatrix} 1 & 1 \\ q & 1 \end{bmatrix}.$$

It is analytic, even linear, in q. Yet its eigenvalues, which are $1\pm\sqrt{q}$, are not analytic.

In general, the eigenvalues of an analytic matrix are not analytic. However, if the unperturbed eigenvalue is semisimple, then at least the lowest order term of the perturbed eigenvalue is linear in the perturbation parameter q.

We shall in the remainder of this section proceed as follows. First, we define, in Section 2.1, semisimple eigenvalues and discuss some useful properties associated with them. Then, in Section 2.2, we discuss the existence of linear lowest order perturbations to semi-simple eigenvalues. Finally, in Section 2.3, we derive formulae for the linear parts of the perturbations of semisimple eigenvalues.

A.2.1. Semisimple eigenvalues.

Let A be an N×N complex matrix. Recall that a complex number λ is an <u>eigenvalue</u> of A if there exists a nonzero N-dimensional complex column vector v such that

(A.11) $Av = \lambda v.$

Then v is an <u>eigenvector</u> of A belonging to the eigenvalue λ.

Writing this as $(A-\lambda)v=0$, we see that if λ is an eigenvalue of A, then the matrix A-λ is singular: it maps a nonzero vector into zero. Therefore it must have zero determinant:

(A.12) $\det(A-\lambda) = 0.$

The left hand side of this equation is a polynomial of degree N in λ, with complex coefficients. It is an immediate consequence of the Fundamental Theorem of Algebra that such a polynomial has exactly N linear factors. So equation (12) can be uniquely written in the form

$$\beta\,(\lambda-\lambda_1)^{m_1}(\lambda-\lambda_2)^{m_2}\,\cdots\,(\lambda-\lambda_s)^{m_s}=0,$$

where β, λ_1, λ_2,..., λ_s are complex numbers are m_1,...,m_s are positive integers. The numbers λ_I, $I=1,...,s$, are the eigenvalues of A, and m_I is the <u>algebraic multiplicity</u> of the eigenvalue λ_I. These algebraic multiplicities satisfy

(A.13) $$\sum_{I=1}^{s} m_s = N.$$

Of course, $s \le N$.

Now, there must exist at least one eigenvector for each eigenvalue —— this was our starting point. The <u>geometric multiplicity</u> of an eigenvalue is the number of linearly independent eigenvectors belonging to that eigenvalue. It is not, however, necessarily the case that the geometric multiplicity of an eigenvalue is the same as its algebraic multiplicity. (When, in the following, we use the term <u>multiplicity</u> without qualification, we will mean algebraic multiplicity.)

Consider, for example, the matrix

(A.14) $$A = \begin{bmatrix} 1 & 1 \\ 0 & 1 \end{bmatrix}.$$

This has only one eigenvalue, namely 1, with multiplicity 2. Let us solve for the eigenvectors. If v is an eigenvector, then $(A-1)v=0$, which is

$$0 = \begin{bmatrix} 0 & 1 \\ 0 & 0 \end{bmatrix}\begin{bmatrix} v_1 \\ v_2 \end{bmatrix} = \begin{bmatrix} v_2 \\ 0 \end{bmatrix}.$$

So every eigenvector v must have $v_2=0$; it must be of the form

$$v = \begin{bmatrix} v_1 \\ 0 \end{bmatrix},$$

where v_1 is a complex number. But no two such vectors are linearly independent: there is only one linearly independent eigenvector belonging to the eigenvalue 1.

In general, the geometric multiplicity of an eigenvalue is less than or equal to its algebraic multiplicity. This is the motivation for

Definition A.1. Let λ be an eigenvalue, with algebraic multiplicity m, of a complex square matrix A. In case there are m linearly independent eigenvectors belonging to the eigenvalue λ, then λ is called a semisimple eigenvalue of A. If all the eigenvalues of A are semisimple, then A is called a semisimple matrix.

Proofs of the following propositions can be found in any good book on linear algebra (for instance, Halmos 1958, Theorem 58.2 and §15, where they are stated in terms of abstract linear operators rather than matrices).

Proposition A.2. Let V be the linear space of all N-dimensional complex column vectors. If λ is a semisimple eigenvalue of an N×N complex matrix A and V_λ is the linear space of all eigenvectors of A belonging to this eigenvalue, then there exists a linear subspace V_C of V such that

(A.15) $V = V_\lambda + V_C$

(direct sum); the subspaces V_λ and V_C are, moreover, both invariant under A.

Proposition A.3. For every basis $\{e_i | \ i=1,\ldots,N\}$ of the linear space V of all N-dimensional complex column vectors, there exists a basis $\{\tilde{e}_i | \ i=1,\ldots,N\}$ of the space $\tilde{V}$ of all N-dimensional complex row vectors such that

$$\tilde{e}_i e_j = \delta_{ij},$$

where δ_{ij} is the Kronecker delta (6). The basis $\{\tilde{e}_i\}$ is called the dual basis to $\{e_i\}$.

We will use these to prove

<u>Proposition A.4</u>. Let λ be a semisimple eigenvalue, with multiplicity m, of a complex square matrix A. Then to each set $\{v_1,\ldots,v_m\}$ of linearly independent (right-) eigenvectors of A belonging to this eigenvalue,

$$Av_I = \lambda v_I , \quad I=1,\ldots,m,$$

there corresponds a set $\{\tilde{v}_1,\ldots,\tilde{v}_m\}$ of linearly independent left-eigenvectors of A belonging to this eigenvalue,

$$\tilde{v}_I A = \lambda \tilde{v}_I , \quad I=1,\ldots,m,$$

satisfying, moreover, the relation

(A.16) $\quad \tilde{v}_I v_J = \delta_{IJ} , \quad I,J=1,\ldots,m,$

where δ_{IJ} is the Kronecker delta (6).

<u>Caveat</u>. In general the row matrix $\tilde{v}_I$ is <u>not</u> the transpose of the column matrix v_I. It is the transpose if and only if A is symmetric.

<u>Proof</u>. Let A, λ, $\{v_1,\ldots,v_m\}$ be as in the hypothesis. Let V_C be the subspace of V defined by the decomposition in Proposition 2, and let $\{v_{m+1},\ldots,v_N\}$ be a basis for V_C. Then $\{v_1,\ldots,v_N\}$ is a basis for V.

Let $\{\tilde{v}_1,\ldots,\tilde{v}_N\}$ be the corresponding dual basis for $\tilde{V}$ (Proposition 3). Then there exist N^2 complex numbers B_{ij} such that

(A.17) $\quad \tilde{v}_i A = \sum_{j=1}^{N} B_{ij}\tilde{v}_j , \quad i=1,\ldots,N.$

For any i, $k\in\{1,\ldots,N\}$,

(A.18) $\quad (\tilde{v}_i A)v_k = \tilde{v}_i(Av_k),$

since matrix multiplication is associative. The left hand side is

(A.19) $(\sum_j B_{ij}\tilde{v}_j)v_k = B_{ik}.$

For i, $k \in \{1,\ldots,m\}$, the right hand side of (18) is

(A.20) $\tilde{v}_i(Av_k) = \tilde{v}_i(\lambda v_k) = \lambda\delta_{ik}.$

Suppose now $i\in\{1,\ldots,m\}$, $k\in\{m+1,\ldots,N\}$. Then since V_C is invariant under A, there exist $(N-m)^2$ complex numbers C_{kh}, k, $h\in\{m+1,\ldots,N\}$, such that

$$Av_k = \sum_{h=m+1}^{N} C_{kh}v_h \ , \quad k=m+1,\ldots,N.$$

Therefore

(A.21) $\tilde{v}_i(Av_k) = \tilde{v}_i(\sum_{h=m+1}^{N} C_{kh}v_h) = 0 \ , \quad i=1,\ldots,m \ , \quad k=m+1,\ldots,N.$

Collecting equations (18)-(21) together, we have finally

$$\tilde{v}_I A = \lambda\tilde{v}_I \ , \quad i=1,\ldots,m.$$

The relation (16) follows from our choice of the basis $\{\tilde{v}_i\}$ as dual to $\{v_i\}$. ∎

A.2.2. Existence of linear perturbations.

We have seen that the perturbed stability matrix A(q) defined by equation (7) is an analytic function of q, and we have also seen that this does not necessarily imply that its eigenvalues are analytic

functions of q. To see what it does imply, consider the characteristic
equation

$$\det(A(q) - \lambda(q)) = 0.$$

This gives the eigenvalues $\lambda(q)$ as the roots of a polynomial of finite
degree, with coefficients which are analytic functions of q. It
follows that the eigenvalues are algebraic functions of q (see, for
instance, Ahlfors 1953), that is, they are branches of analytic functions with, in general, algebraic singularities.

However, if the unperturbed eigenvalue is semisimple, the behavior of $\lambda(q)$ is more restricted:

Proposition A.5. Let A(q) be a complex square matrix which
depends analytically on a complex parameter q, and suppose $\lambda^{(0)}$ is a
semisimple eigenvalue, with multiplicity m, of A(0). Then A(q) has
m eigenvalues of the form

$$(A.22) \qquad \lambda(q) = \lambda^{(0)} + \lambda_j^{(1)} q + q^{(1+1/r_j)} (\beta_{jk}^{(0)} + \beta_{jk}^{(1)} q + \cdots),$$

where $\lambda_j^{(1)}$, $\beta_{jk}^{(0)}$, $\beta_{jk}^{(1)}$, $\cdots$ are complex numbers and r_j are integers
equal to or greater than 1. For each j there are m_j such eigenvalues,
with $m_j = m$, and $k = 1, \ldots, m_j$.

Proof. Baumgärtel (1964); Kato (1966). ∎

This proposition says that if an unperturbed eigenvalue is semisimple,
then the leading term of each corresponding perturbed eigenvalue is
linear in q.

For instance, the matrix

$$\begin{bmatrix} 1 & q \\ q & 1 \end{bmatrix}$$

has 1 as its only unperturbed eigenvalue, and this eigenvalue is semisimple with multiplicity 2. The corresponding perturbed eigenvalues, which are 1±q, are indeed of the form (22). On the other hand, the matrices (10) and (14) provide an example of a non-semisimple unperturbed eigenvalue which has perturbed eigenvalues whose leading terms are not linear.

A.2.3. Calculation of linear perturbations.

Our object in this section is to find formulae with which we can calculate the coefficients $\lambda_j^{(1)}$ of the first-order corrections to semisimple eigenvalues due to analytic perturbations (equation (22)). Similar formulae to the ones we will arrive at are well-known among physicists (see any textbook on quantum mechanics), but apply only to self-adjoint perturbations of self-adjoint matrices; such a restriction is far too strong for our purposes.

We first need to know how the perturbed eigenvectors behave. This is given by

Lemma A.1. Let $A(q)$ be an N×N complex matrix, analytic in q. Then for each eigenvalue (not necessarily semisimple) of $A(q)$, there exists an eigenvector of the form

$$(A.23) \qquad v(q) = v^{(0)} + v^{(1)}q + \sum_{k=1}^{r-1} w^{(k)} q^{k/r} + o(q),$$

where $v^{(0)}$ is an eigenvector of $A(0)$, r is an integer greater than 1, and $v^{(1)}$ and $w^{(k)}$, k=1,...,r-1, are constant vectors.

Remark. The notation $f(q)=g(q)+o(q)$ means that

$$\lim_{q \to 0} \frac{f(q)-g(q)}{q} = 0.$$

This is not to be confused with the other notation $f(q)=g(q)+O(q)$, which means that $(f-g)/q$ is merely bounded in a neighborhood of q=0.

Proof of Lemma 1. Each eigenvector $v(q)$ satisfies $A(q)v(q)=$ $=\lambda(q)v(q)$, which is a system of N homogeneous equations

$$(A_{11}-\lambda)v_1+A_{12}v_2+\ \cdot\ \cdot\ \cdot\ +A_{1N}v_N=0$$

$$\cdot$$

$$\cdot$$

$$A_{N1}v_1+\ \cdot\ \cdot\ \cdot\quad\ \cdot\ +(A_{NN}-\lambda)v_N=0$$

for the components v_i. The solutions can be obtained by a finite number of rational algebraic operations, and are therefore expressible as
polynomials in the coefficients $(A_{11}-\lambda)$, $A_{12},\ldots,(A_{NN}-\lambda)$. Let $v_i=u_i$
be such a solution. Recall that the A_{ij} are analytic in q, and that
λ is an algebraic function of q so that it can be expressed as a series
of positive powers of $q^{1/r}$ for some integer $r\geqslant 1$.

If any one of the u_i is nonzero for q=0, then $v_i=u_i$ is an eigenvector of the required form.

If all u_i are zero at q=0, let s_i be the order of the zero of u_i
at q=0. Let s be the smallest of the s_i. Then $v_i=q^{-s}u_i$ is an eigenvector of the required form.

This says that each perturbed eigenvector continuously approaches
an unperturbed eigenvector. One might be tempted to suppose that if
the unperturbed eigenvalue is semisimple with multiplicity m, so that
there are m linearly independent unperturbed eigenvectors, then there
would be m linearly independent perturbed eigenvectors which continuously approach the m unperturbed eigenvectors. A counterexample to
this is provided by the matrix

$$\begin{bmatrix}1 & q \\ 0 & 1\end{bmatrix},$$

which has two linearly independent unperturbed eigenvectors but only
one linearly independent perturbed eigenvector. We therefore have to
proceed a little indirectly, first proving

Lemma A.2. For any square complex matrix $A(q)$ which is analytic in q, there exists a nonsingular matrix $S(q)$ such that

$$A'(q) = S^{-1}(q)A(q)S(q) = \begin{bmatrix} \lambda_1(q) & A'_{12}(q) \cdots & \cdot & A'_{1N}(q) \\ \\ 0 & \lambda_2(q) & \cdots & \cdot & A'_{2N}(q) \\ \\ & & \cdot \\ \\ & & \cdot \\ \\ 0 \cdots & & \cdot & 0 & \lambda_N(q) \end{bmatrix}$$

where $\lambda_1(q),\ldots,\lambda_N(q)$ are the eigenvalues of $A(q)$, and the elements of $S(q)$ have the form

(A.24) $S_{ij}(q) = S_{ij}^{(0)}+S_{ij}^{(1)}q+F(q)+o(q),$

where $F(q)$ denotes a finite sum of fractional powers of q, the powers lying between 0 and 1.

Moreover, if $\lambda_i(0)$ is a semisimple eigenvalue of $A(0)$, then the off-diagonal elements $A'_{ij}(q)$ with $j>i$ are zero.

Proof. This is just a special case of the Jordan canonical form. The additional restrictions here can easily be obtained in the course of any standard proof of the existence of the Jordan canonical form, using the extra information in Proposition 5 and Lemma 1.

Theorem A.1. Let

$$A(q) = A^{(0)} + A^{(1)}q + \cdots$$

be a complex N×N matrix, analytic in q, and let $\lambda^{(0)}$ be a semisimple eigenvalue of $A^{(0)}$, with multiplicity m. Let $\{u_I^{(0)}\mid I=1,\ldots,m\}$ be m linearly independent (right-) eigenvectors (column matrices) of $A^{(0)}$ belonging to the eigenvalue $\lambda^{(0)}$, and $\{\tilde{u}_I^{(0)}\mid I=1,\ldots,m\}$ the m corresponding left-eigenvectors (row matrices) which satisfy $\tilde{u}_I u_J = \delta_{IJ}$ (Proposition 4).

Then m eigenvalues of A(q) have the form

$$\lambda_I(q) = \lambda^{(0)} + \lambda_I^{(1)} q + o(q),$$

where the complex numbers $\lambda_I^{(1)}$, I=1,...,m, are the eigenvalues of the m×m **perturbation matrix** P with elements

(A.25) $\qquad P_{IJ} = \tilde{u}_I^{(0)} A^{(1)} u_J^{(0)}, \quad I,J=1,...,m.$

Proof. Since $\lambda^{(0)}$ has multiplicity m it occurs as the zero-order part of m entries along the main diagonal of the semidiagonalized matrix A'(q) of Lemma 2. For the sake of convenience we can take these to be the first m entries along the main diagonal. Define m column vectors v_I', I=1,...,m, by

$$(v_I')_j = \delta_{Ij}, \quad I=1,...,m, \quad j=1,...,N,$$

where $(v_I')_j$ denotes the jth entry from the top of the column matrix v_I'. Then from the semidiagonal form of Lemma 2,

$$S^{-1}(q)A(q)S(q)v_I' = A'(q)v_I' = \lambda_I(q)v_I' + \sum_{J<I} A_{JI}'(q)v_J'.$$

Multiplying both sides of this equation by S(q), we get

(A.26) $\qquad A(q)v_I(q) = \lambda_I(q)v_I(q) + \sum_{J<I} A_{JI}'(q)v_J(q), \quad I=1,...,m,$

where

$$v_I(q) = S(q)v_I'$$

whence

$$(v_I(q))_j = S_{jI}(q).$$

Because of the form (24) of $S(q)$, each $v_I(q)$ has the form

$$v_I(q) = v_I^{(0)} + v_I^{(1)}q + F(q) + o(q)$$

where $v_I^{(0)}$ and $v_I^{(1)}$, $I=1,\ldots,m$, are constant vectors. Since $A'_{JI}(q)=0$ for $I>J$, (26) implies that each $v_I^{(0)}$ is an eigenvector of $A^{(0)}$ with eigenvalue $\lambda^{(0)}$.

Let $\{u_I^{(0)} \mid I=1,\ldots,m\}$ be any m linearly independent eigenvectors of $A^{(0)}$ belonging to the eigenvalue $\lambda^{(0)}$. Then there exist constants C_{IJ}, $I,J=1,\ldots,m$, such that

$$v_I^{(0)} = \sum_J C_{IJ} u_J^{(0)}.$$

We can, then, write equation (26) as

$$(A^{(0)} + A^{(1)}q + o(q))(\sum_J C_{IJ} u_J^{(0)} + v_I^{(1)}q + F(q) + o(q))$$

(A.27)
$$= (\lambda^{(0)} + \lambda_I^{(1)}q + o(q))(\sum_J C_{IJ} u_J^{(0)} + v_I^{(1)}q + F(q) + o(q)) +$$

$$+ \sum_{J<I} A'_{JI}(q)(\sum_K C_{JK} u_K^{(0)} + v_J^{(1)}q + F(q) + o(q)).$$

The terms of order zero in q are just the eigenvalue equation for $A^{(0)}$:

$$A^{(0)} \sum_J C_{IJ} u_J^{(0)} = \lambda^{(0)} \sum_J C_{IJ} u_J^{(0)},$$

which holds identically.

For I=1, the terms of first order in q are

$$A^{(1)} \sum_J C_{1J} u_J^{(0)} + A^{(0)} v_1^{(1)} = \lambda^{(0)} v_1^{(1)} + \lambda_1^{(1)} \sum_J C_{1J} u_j^{(0)}.$$

But we need not have taken λ_1 as the first eigenvalue along the main diagonal in the semidiagonal form; we could just as well have taken any of the λ_I. We would then have found in the same way for, say, λ_I the equation

$$(A.28) \qquad A^{(0)} v_I^{(1)} + A^{(1)} \sum_J C_{IJ} u_J^{(0)} = \lambda^{(0)} v_I^{(1)} + \lambda_I^{(1)} \sum_J C_{IJ} u_J^{(0)}.$$

Now let $\{\tilde{u}_I^{(0)} \mid I=1,\ldots,m\}$ be the left-eigenvectors of $A^{(0)}$ belonging to the eigenvalue $\lambda^{(0)}$ and satisfying

$$\tilde{u}_I u_J = \delta_{IJ}, \qquad I,J=1,\ldots,m.$$

(Their existence is guaranteed by Proposition 4.) Multiplying both sides of equation (28) from the left with $\tilde{u}_K^{(0)}$ and remembering that $\tilde{u}_K^{(0)}$ is a left-eigenvector of $A^{(0)}$, we get

$$\lambda^{(0)} \tilde{u}_K^{(0)} v_I^{(1)} + \sum_J \tilde{u}_K^{(0)} A^{(1)} u_J^{(0)} C_{IJ} = \lambda^{(0)} \tilde{u}_K^{(0)} v_I^{(1)} + \sum_J \lambda_I^{(1)} \delta_{KJ} C_{IJ},$$

that is,

$$\sum_J (\tilde{u}_K^{(0)} A^{(1)} u_J^{(0)} - \lambda_I^{(1)} \delta_{KJ}) C_{IJ} = 0, \qquad I=1,\ldots,m.$$

But this says precisely that the coefficients $\lambda_I^{(1)}$, $I=1,\ldots,m$, are the m eigenvalues of the perturbation matrix defined in the statement of the theorem.

A.3. Calculation of Λ.

Local stability of an equilibrium can conveniently be expressed by the parameter Λ defined in Section II.2. We will in this section calculate Λ for colonization equilibria of founder controlled communities (Chapter II) and for reshuffled equilibria of dominance controlled communities (Chapter III).

A.3.1. Colonization equilibria.

We consdier first the colonization equilibria defined in Section II.1. Our system of differential equations is (I.6), which we can write in the form

$$(A.29) \qquad \frac{dx_i^\mu}{dt} = F_i^\mu(x;D) = F^{(0)\mu}_i(x) + F^{(1)\mu}_i(x)D, \quad \mu=1,\ldots,m, \quad i=1,\ldots,n,$$

with

$$(A.30) \qquad F^{(0)\mu}_i(x) = C_i x_i^\mu (1 - \sum_j \alpha_{ij} x_j^\mu)$$

$$(A.31) \qquad F^{(1)\mu}_i(x) = \sum_\nu E^{\mu\nu} x_i^\nu.$$

Notice that the single index i of the preceding sections of this chapter is now replaced by the <u>pair</u> of indices $\genfrac{}{}{0pt}{}{\mu}{i}$.

The unperturbed equilibrium is given by the colonization initial values

$$(A.32) \qquad \hat{x}_i^\mu = \delta_{ic(\mu)}.$$

The unperturbed stability matrix is

$$A^{(0)\mu\nu}_{ij} = \frac{\partial F^{(0)\mu}_i}{\partial x_j^\nu}(\hat{x}),$$

which we can write as

$$A^{(0)\mu\nu}_{ij} = \delta_{\mu\nu}A_{ij}(c(\mu)),$$

where $A_{ij}(c(\mu))$ is the stability matrix for the Lotka-Volterra equilibrium with one species, $c(\mu)$, present (Section B.1).

We proceed to find the eigenvalues and eigenvectors of the unperturbed stability matrix. Let $\{v_i(c(\mu))\}$ be the (right-) eigenvectors of the single-species Lotka-Volterra stability matrix $A_{ij}(c(\mu))$ (equations (B.6)):

$$\sum_j A_{ij}(c(\mu))(v_k(c(\mu)))_j = \lambda_k(c(\mu))(v_k(c(\mu)))_i.$$

Then the m×n-dimensional column vectors $\{v_i^\mu| \mu=1,\ldots,m, i=1,\ldots,n\}$ defined by the components

(A.33) $\quad (v_i^\mu)_j^\nu = \delta_{\mu\nu}(v_i(c(\mu)))_j$

are eigenvectors of $A^{(0)}$:

$$\sum_{\nu,j} A^{(0)\mu\nu}_{ij}(v_k^\sigma)_j^\nu = \sum_{\nu,j} \delta_{\mu\nu}A_{ij}(c(\mu))\delta_{\sigma\nu}(v_k(c(\sigma)))_j =$$

$$= \delta_{\mu\sigma}\sum_j A_{ij}(c(\sigma))(v_k(c(\sigma)))_j = \lambda_k(c(\sigma))\delta_{\mu\sigma}(v_k(c(\sigma)))_i =$$

$$= \lambda_k(c(\sigma))(v_k^\sigma)_i^\mu,$$

with eigenvalues

$$(A.34) \quad \begin{cases} \lambda^{(0)\mu}_i = \lambda_i(c(\mu)) = -C_i(\alpha_{ic(\mu)}{}^{-1}), & i=c(\mu) \\[2mm] \lambda^{(0)\mu}_{c(\mu)} = \lambda_{c(\mu)}(c(\mu)) = -C_{c(\mu)}. \end{cases}$$

Notice that the m×n eigenvectors defined by (33) are linearly independent: the (m×n)×(m×n) matrix $A^{(0)}$ has m×n linearly independent eigenvectors, which implies that the matrix is semisimple, that is, every
eigenvalue is semisimple.

In the same way, one can show that the m×n-dimensional row vector
$\tilde{v}_i^\mu$ with components

(A.35) $$(\tilde{v}_i^\mu)_j^\nu = \delta_{\mu\nu}(\tilde{v}_i(c(\mu)))_j,$$

where $\{\tilde{v}_i(c(\mu))\}$ are the left-eigenvectors of the single-species Lotka-
Volterra stability matrix $A_{ij}(c(\mu))$, is a left-eigenvector of $A^{(0)}$ belonging to the eigenvalue $\lambda^{(0)}{}_i^\mu$, and that

$$\tilde{v}_i^\mu v_j^\nu = \delta_{\mu\nu}\delta_{ij}.$$

One easily sees that the inverse matrix to $A^{(0)}$ has the elements

(A.36) $$(A^{(0)-1})_{ij}^{\mu\nu} = \delta_{\mu\nu}A_{ij}^{-1}(c(\mu)),$$

where the Lotka-Volterra matrix $A^{-1}(c(\mu))$ is given by equations (B.4)
and (B.5). Therefore, the first-order perturbation $A^{(1)}$ to the stability matrix (equation (9)) is

$$A^{(1)\mu\nu}_{ij} = -\sum_{\substack{\rho,\sigma \\ s,t}} \frac{\partial^2 F_i^{(0)\mu}}{\partial x_j^\nu \partial x_s^\rho}(\hat{x})(A^{(0)-1})_{st}^{\rho\sigma}F_t^{(1)\sigma}(\hat{x}) + \frac{\partial F_i^{(1)\mu}}{\partial x_j^\nu}(\hat{x})$$

(A.37) $$= \delta_{ij}E_i^{\mu\nu} + \delta_{\mu\nu}\sum_{\xi,s} C_i(\delta_{ij}\alpha_{is} + \delta_{is}\alpha_{ij}) \times$$

$$\times (\alpha_{sc(\xi)}T_s(c(\mu)) + \delta_{sc(\mu)}U_{c(\xi)}(c(\mu)))E_{c(\xi)}^{\mu\xi},$$

where we have used equations (B.4) and (B.10), and the quantities T and U are given by equations (B.5).

It will save pointless complication if we assume that all n^2 eigenvalues $\lambda_i(j)$, $i,j=1,\ldots,n$, (equation (B.8)) at our n single-species Lotka-Volterra equilibria are distinct. This will certainly hold in the generic case; that is, if it fails to hold for any system it can be made to hold by an arbitrarily small adjustment of the parameters C_i and α_{ij}.

Our object is to calculate Λ for the perturbed equilibrium. Recall that Λ is the negati e of the largest eigenvalue (equation (II.6); all these eigenvalues are real). We will calculate it by calculating the largest perturbation to the largest unperturbed eigenvalue; this will be correct if D is small enough. We have shown that all eigenvalues of $A^{(0)}$ are semisimple. Therefore, we can calculate the first-order perturbations by using Theorem 1 of Section 2.3.

Let $\Lambda^{(0)}=-\text{maximum}\{\lambda^{(0)\mu}_i\}$, where $\{\lambda^{(0)\mu}_i\}$ are the unperturbed eigenvalues (34). Then for all $\lambda^{(0)\mu}_i=-\Lambda^{(0)}$, because the Lotka-Volterra eigenvalues $\lambda_i(j)$ are all distinct: 1) i has always the same value, 2) but μ will generally take on several different values, namely all those for which $c(\mu)$ has its fixed value. Let us say that

(A.38) $i=a$, $c(\mu)=b$ for all $\lambda^\mu_i=-\Lambda^{(0)}$.

Let m_b be the number of patches μ for which $c(\mu)=b$. Then m_b is the multiplicity of the unperturbed eigenvalue $-\Lambda^{(0)}$.

Let μ_1, $\mu_2,\ldots$, μ_{m_b} be the values of μ such that $c(\mu)=b$. Then the perturbation matrix (25) for the eigenvalue $-\Lambda^{(0)}$ is given by

(A.39) $P_{IJ} = \tilde{v}^{\mu_I}_a A^{(1)} v^{\mu_J}_a$,

where $A^{(1)}$ is given by equation (37), the column vector $v^{\mu_I}_a$ by (33), and the row vector $\tilde{v}^{\mu_I}_a$ by (35). Thus, according to Theorem 1, the m_b perturbed eigenvalues corresponding to the unperturbed eigenvalue $-\Lambda^{(0)}$ are, to first order in D,

$$\{-\Lambda^{(0)} + \lambda_I^{(1)} D \mid I=1,\ldots,m_b\} \;,$$

where $\lambda_I^{(1)}$, $I=1,\ldots,m_b$, are the eigenvalues of the perturbation matrix (39). Finally, Λ is minus the largest of these (because they are all real):

(A.40) $\qquad \Lambda = \Lambda^{(0)} - \text{maximum}\{\lambda_I^{(1)}\}D.$

We find with a little labor that

(A.41) $\qquad P_{IJ} = E_a^{\mu_I \mu_J} + \delta_{\mu_I \mu_J} \sum_{\sigma=1}^{m} X(\sigma) E_{c(\sigma)}^{\mu_I \sigma} \;,$

where $X(\sigma)$ is given by the following formulae. If $a=b$ in equation (38)

(A.42) $\qquad X(\sigma) = \begin{cases} \dfrac{C_b \alpha_{bc(\sigma)}}{C_{c(\sigma)} \left(\alpha_{c(\sigma)b} - 1\right)} \quad 1 - \dfrac{C_{c(\sigma)} \alpha_{c(\sigma)b}}{C_b - C_{c(\sigma)} \left(\alpha_{c(\sigma)b} - 1\right)} \quad , \; c(\sigma) \neq b \\[4ex] -2 \quad , \quad c(\sigma) = b. \end{cases}$

If $a \neq b$,

(A.43) $\qquad X(\sigma) = \begin{cases} \dfrac{C_a}{C_{c(\sigma)}} \; \dfrac{\alpha_{ab} \alpha_{bc(\sigma)} - \alpha_{ac(\sigma)}}{\alpha_{c(\sigma)b} - 1} \quad , \; c(\sigma) \neq a, \; c(\sigma) \neq b \\[4ex] \dfrac{1}{\alpha_{ab} - 1} \; \alpha_{ab} \alpha_{ba} \; \dfrac{2C_b - C_a(\alpha_{ab} - 1)}{C_b - C_a(\alpha_{ab} - 1)} - 2 \quad , \; c(\sigma) = a \\[4ex] -\dfrac{C_a \alpha_{ab}}{C_b} \quad , \quad c(\sigma) = b. \end{cases}$

The form of the perturbation matrix (41) depends on our choice for the dispersal coefficients $E^{\mu\nu}$. We complete the calculation of Λ by showing how to find the largest eigenvalue of P for each of our two forms for the dispersal coefficients (Section I.4, item 3)).

*

If the dispersal coefficients have the form (I.7) characterizing random dispersal, then the perturbation matrix (41) has the form

$$
P = \begin{bmatrix}
X & 1 & 1 & \cdot & \cdot & & \cdot & 1 \\
1 & X & 1 & \cdot & \cdot & & \cdot & 1 \\
 & & & \cdot & & & & \\
 & & & \cdot & & & & \\
 & & & \cdot & & & & \\
1 & 1 & \cdot & \cdot & \cdot & & \cdot & 1 & X
\end{bmatrix}
$$

That is, all off-diagonal entries are 1 and each diagonal entry is equal to

(A.44) $X = \displaystyle\sum_{\sigma:c(\sigma)=b} X(\sigma)+(m_b-m)X(\mu_b)+1-m$,

where μ_b is any patch for which $c(\mu_b)=b$. The eigenvalues of matrices of this form are well known; they are discussed, for example, in Appendix II of May's book (1974). The largest eigenvalue is

(A.45) $\max\{\lambda_I^{(1)}\} = X+m_b-1$,

and using this in equation (40) completes the calculation of Λ for this case.

*

In the case of linear nearest neighbor dispersal, the dispersal coefficients have the form (I.8) and P has the form

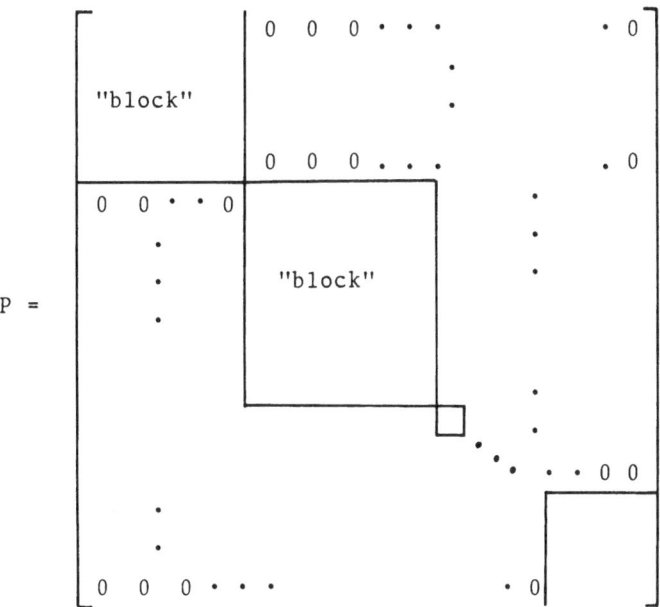

$$P = \begin{bmatrix} \text{"block"} & \begin{matrix} 0 & 0 & 0 & \cdots & & & \cdot\ 0 \\ & & & & & & \\ 0 & 0 & 0 & \cdots & & & \cdot\ 0 \end{matrix} \\ \begin{matrix} 0 & 0 & \cdots & \cdot\ 0 \\ & \vdots & & \\ & \vdots & & \end{matrix} & \text{"block"} \\ & & \cdot\cdot\ 0\ 0 \\ \begin{matrix} 0 & 0 & 0 & \cdots \end{matrix} & \cdot\ 0 \end{bmatrix}$$

That is: all entries outside a sequence of square "blocks" along the
main diagonal are zero, and each "block" has the form

$$\begin{bmatrix} B_1 & 1 & 0 & 0 & \cdot & \cdot & & \cdot\ 0 \\ 1 & B_2 & 1 & 0 & \cdot & \cdot & & \cdot\ 0 \\ & & & \cdot & & & & \\ & & & \cdot & & & & \\ & & & \cdot & & & & \\ & & & & & \cdot & & \\ 0 & 0 & \cdot & \cdot & & \cdot & 0 & 1 & B_M \end{bmatrix}$$

Each block corresponds to a row of M adjacent patches, with μ
equal to, say, ν, $\nu+1,\ldots$, $\nu+M$, such that $c(\mu)=b$ for all μ $\{\nu,\ldots,\nu+M\}$
but $c(\nu-1)=b$ and $c(\nu+M+1)=b$. Then the block is an M×M matrix, with
the following entries along the main diagonal:

$$B_1 = X(\nu-1)-X(\nu)-2$$

$$B_I = -2 , \quad 1<I<M$$

$$B_M = X(\nu+M+1)-X(\nu+M)-2.$$

(For the two patches at the extreme ends of the whole row of m patches obvious corrections have to be made.) Each entry adjacent to the main diagonal is 1, and all remaining entries are 0. For each isolated patch ν with $c(\nu)=b$, $c(\nu-1)=b$, and $c(\nu+1)=b$, the block is simply the single entry (1-dimensional block) $X(\nu-1)+X(\nu+1)-2X(\nu)-2$ in the main diagonal of P.

Such a perturbation matrix P is reducible: its eigenvalues are just the eigenvalues of all the blocks that appear in it. Unfortunate-ly, it appears to be impossible to express the eigenvalues of the blocks in closed form. However, they can be computed numerically by means of the algorithm of Givens for triple-diagonal matrices, which is described, for example, by Fox (1964, pp. 243-245). With this al-gorithm one has excellent control over the convergence of the approx-imation, and one can choose to calculate the kth largest eigenvalue only, for any k; we need only the largest eigenvalue.

Thus a numerical calculation of Λ proceeds as follows. First the function $c(\mu)$ is scanned to find all the "blocks" in P; then the lar-gest eigenvalue for each block is computed by Givens' algorithm; and finally the largest of these is used in equation (40).

*

In order to evaluate Λ numerically, we have to choose numerical values for the Lotka-Volterra parameters C_i and α_{ij}. But it is not very informative to tie ourselves down to any one choice for these parameters, and we would like moreover to be able to meaningfully compare systems with different numbers of species. We will, therefore, average Λ over suitably chosen statistical ensembles of systems.

We define such an ensemble as follows.

1) The number m of patches is fixed throughout these calculations at 200.

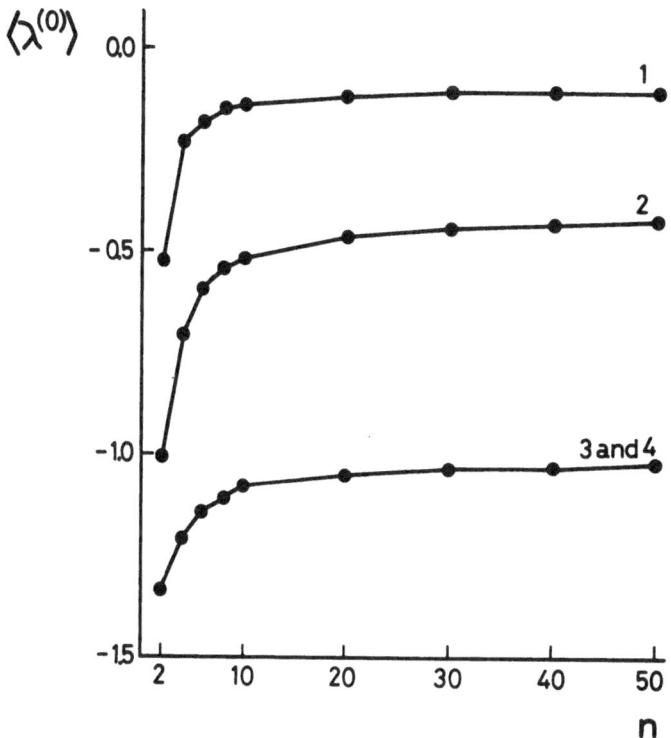

Figure A.1. The average (over the statistical
ensemble defined in the text) largest unperturbed
eigenvalue, as a function of the number n of colonizing species, for each of the four choices
specified in the text for the interval (e,f) from
which the competition coefficients are chosen.

2) The parameters C_i are randomly distributed over the interval
(1,2) of real numbers, with uniform probability distribution.

3) the parameters α_{ij} with i=j are randomly distributed over the
positive real interval (e,f) with uniform probability distribution.

4) The n species are initially distributed at random over the m
patches, with one species per patch (colonization initial values,
Section I.5) according to one of the two random processes described
in Section I.5.

We denote the average of Λ over such an ensemble by Λ .

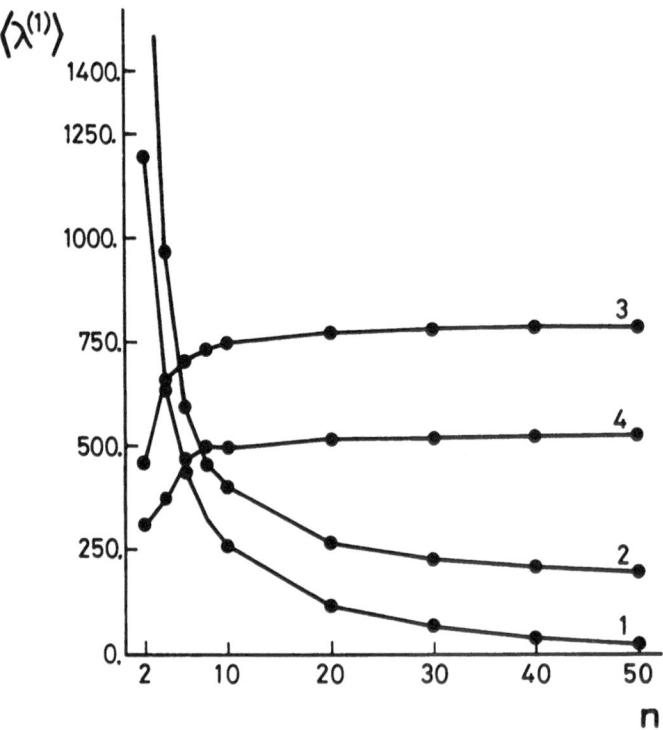

Figure A.2. The average (over the statistical ensemble defined in the text) largest perturbation of the largest eigenvalue, as a function of the number n of colonizing species, for the case of random dispersal. This relation is plotted for each of the four choices specified in the text for the interval (e,f) from which the competition coefficients are chosen.

This quantity was computed by a Monte Carlo method for the following four choices for the interval (e,f):

Choice 1: (e,f)=(1.1,1.9)

Choice 2: (e,f)=(1.5,2.5)

Choice 3: (e,f)=(3.0,4.0)

Choice 4: (e,f)=(10.0,20.0).

Each average was over a sample of 500 systems, except for choice 4 for (e,f), in which case the sample size was 100.

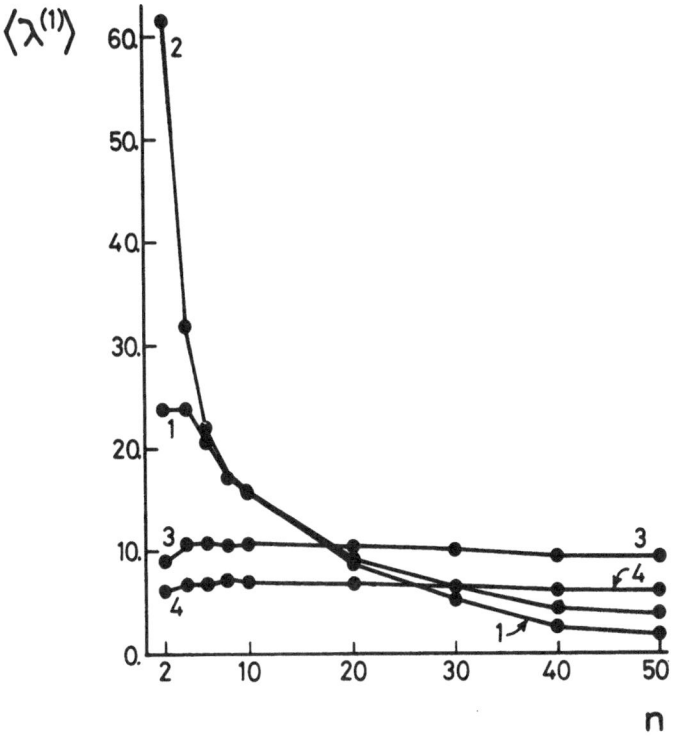

Figure A.3. The average (over the statistical
ensemble defined in the text) largest perturbation of the largest eigenvalue, as a function
of the number n of colonizing species, for the
case of linear nearest neighbor dispersal. This
relation is plotted for each of the four choices
specified in the text for the interval (e,f)
from which the competition coefficients are chosen.

Figures 1-3 show the results of these computations for uniform
colonization (Section I.5). Figure 1 shows the average value of the
largest unperturbed eigenvalue (recall again that the eigenvalues and
their perturbations are real in these calculations) $\lambda^{(0)} = -\Lambda^{(0)}$. This
quantity does not, of course, depend on the form of the dispersal coefficients. Figure 2 shows the average value of the largest perturbation $\lambda^{(1)}$ to this eigenvalue for random dispersal, and figure 3 for
linear nearest neighbor dispersal.

The difference between the qualitative behavior of $\lambda^{(1)}$ for Choices 1 and 2 for the interval (e,f) on the one hand, and for Choices 3 and 4 on the other, is quite striking. This difference reflects the distinction whether a is or is not equal to b in equations (38), through equations (42) and (43).

Suppose first that e>2 (Choices 3 and 4); then we will have for every system in our ensemble a=b, and the X's are given by (42). But since C_b is by definition smaller than $C_i(\alpha_{ib}-1)$ for any i, it follows from (42) that $X(\sigma)>0$ if $c(\sigma)=b$, and $X(\sigma)=-2$ if $c(\sigma)=b$. Now, from equations (44) and (45), for random dispersal,

$$\lambda^{(1)} = \sum_{i=b} m_i X_i - (m-m_b)(X_b+1),$$

where m_i is the number of patches which are colonized by species i, and $X_i=X(\sigma)$ for any σ such that $c(\sigma)=i$. But $\langle X_i \rangle$ is independent of i so long as i=b, and if we set $\langle X_i \rangle=A>0$ for all i=b, we can then write

$$\left\langle \lambda^{(1)} \right\rangle = (n-1)\frac{m}{n}A + (m-\frac{m}{n}) = m(A+1)(1-\frac{1}{n}).$$

We see that $\lambda^{(1)}$ is always positive, and that its mean value is a monotonically increasing function of n, in agreement with the Monte Carlo computation (figure 2).

It follows that $\langle \Lambda \rangle$ is a monotonically decreasing function of n for any value of D which is small enough so that this linear approximation makes sense, for an ensemble with e>2. It might be supposed that this monotonic decrease means that the species richness of these communities is controlled by local stability, but the dependence on n is so weak that any such effect would be far too sensitive to the value of D to be of any significance (see Section II.2).

The situation is more complicated if, for instance, (e+f)/2<2 (Choices 1 and 2), in which case a=b for every system in the ensemble and equation (43) applies. The signs of the X's in (43) are ambiguous, so we cannot easily check the Monte Carlo computations in this case, but according to those computations $\lambda^{(1)}$ tends now to decrease monotonically with n. However, it is questionable whether this is telling us anything about the behavior of Λ, for if some other eigenvalue increases with n it may eventually become the largest eigenvalue, violating our assumption that the largest eigenvalue is given by the largest perturbation to the largest unperturbed eigenvalue. (This is not taken into account by Yodzis 1976b.)

Similar computations were done with colonization by space preemption (Section I.5) for Choices 1 and 3 for (e,f). For each choice of (e,f) and for each number n of colonizing species, the results were essentially the same as for uniform colonization, showing that local stability in these communities is unaffected by numerical dominance (Section III.1) within the community.

A.3.2. Reshuffled equilibria.

In this section we calculate Λ for the reshuffled equilibria defined in Section III.4. The calculation is quite similar to that for colonization equilibria discussed in the preceding section, so we will just sketch it here.

The unperturbed equilibrium is now given by (III.9) as

$$(A.46) \qquad \hat{x}_i = \delta_{is(\mu)} \hat{x}_{s(\mu)}^{(t(\mu))} + {}_{it(\mu)}\hat{x}_{t(\mu)}^{(s(\mu))},$$

where $s(\mu)$ and $t(\mu)$ are obtained from a reshuffle. Recall our conventions (III.7) and (III.8), which enable us to use the two-species formulae of Section B.2 also for those patches with only one predominant species.

The unperturbed stability matrix and its inverse are now given by

$$A^{(0)\mu\nu}_{ij} = \delta_{\xi\nu} A_{ij}(s(\mu),t(\mu))$$

$$(A^{(0)-1})^{\mu\nu}_{ij} = \delta_{\mu\nu} A^{-1}_{ij}(s(\mu),t(\mu)),$$

where the Lotka-Volterra matrices $A(s,t)$ and $A^{-1}(s,t)$ are now given by equations (B.13-18). The eigenvalues of $A^{(0)}$ and the corresponding right- and left-eigenvectors are given by similar, obvious changes to equations (33)-(35), where one now uses equations (B.19) and (B.21-23). The first-order part of the stability matrix is

$$A^{(1)\mu\nu}_{ij} = E^{\mu\nu}_i \delta_{ij} + \delta_{\mu\nu} \sum_{\xi,k} C_i (\delta_{ij}\alpha_{ik} + \delta_{ik}\alpha_{ij}) \times$$

$$\times ((\delta_{ks(\xi)} U_k(s(\mu),t(\mu)) + \delta_{ks(\mu)} V_{s(\xi)}(s(\mu),t(\mu)) +$$

(A.47)
$$+\delta_{kt(\mu)} W_{s(\xi)}(s(\mu),t(\mu))) E_{s(\xi)} \hat{x}_{s(\xi)}(t(\xi)) +$$

$$(\delta_{kt(\xi)} U_k(s(\mu),t(\mu)) + \delta_{kt(\mu)} V_{t(\xi)}(s(\mu),t(\mu)) +$$

$$+\delta_{kt(\mu)} W_{t(\xi)}(s(\mu),t(\mu))) E_{t(\xi)} \hat{x}_{t(\xi)}(s(\xi)).$$

We are again interested in perturbations of the eigenvalue $-\Lambda^{(0)}=\text{maximum}\{\lambda^{(0)\mu}_i\}$. In the present case, the set of all $\lambda^{(0)\mu}_i = -\Lambda^{(0)}$ can be characterized as those $\lambda^{(0)\mu}_i$ for which

$$i=a , \quad \{s(\mu),t(\mu)\}=\{b,c\}$$

for some integers a, b, c. The multiplicity $m_{\{b,c\}}$ of the unperturbed eigenvalue $-\Lambda^{(0)}$ is now the number of patches for which $\{s(\mu),t(\mu)\}= =\{b,c\}$ (unordered sets!).

Let $\{\mu_I|$ $I=1,\ldots,m_{\{b,c\}}\}$ be those patches for which $\{s(\mu),t(\mu)\}= =\{b,c\}$. We now find for the perturbation matrix

$$P_{IJ} = E^{\mu_I\mu_J}_a + \delta_{IJ} \sum_{\sigma} (Y(\sigma) E^{\mu_I\sigma}_{s(\sigma)} + Z(\sigma) E^{\mu_I\sigma}_{t(\sigma)})$$

where Y and Z are obtained by substitution of the appropriate quanti- ties from Section B.2 in equation (47), and multiplication by the eigenvectors as in the preceding section. The explicit expressions for Y and Z are extremely complicated and we will not reproduce them here.

The remainder of the calculation proceeds exactly as in the preceding section, with $X(\sigma)$ replaced everywhere by $Y(\sigma)+Z(\sigma)$ (because we have chosen the dispersal coefficients the same for all species), and with the condition $c(\mu)=b$ replaced by $\{s(\mu),t(\mu)\}=\{b,c\}$.

*

The average value of Λ over statistical ensembles defined as in the preceding section has been computed by Monte Carlo methods, again as in Section 3.1. However, in these calculations we have always taken $e=0.5$, and allowed f to vary so as to give a range of values of the parameter $p=(1-e)/(f-e)$. Because the species richness of these reshuffled communities is governed primarily by dominance relations rather than by the number of colonizing species (Section III.4.2), it is no longer possible to specify independently the interval (e,f) and the number of species in the community.

Therefore, for each choice of (e,f) only one choice for the number n of colonizing species was used, this number being chosen so that saturation in species packing was being approached (Section III.4.2). The results of these Monte Carlo computations are given in table 1, in which the first column gives the chosen value for f, the second

Table A.1

f	p	Number of colonizing species	Average S	$\left\langle \lambda^{(0)} \right\rangle$	$\left\langle \lambda^{(1)} \right\rangle$
9.0	.059	20	7.80	-0.077	14607.
13.5	.038	20	10.40	-0.100	7613.
19.0	.027	30	14.38	-0.062	17128.
25.5	.020	40	18.96	-0.049	28063.
33.0	.015	40	22.16	-0.052	32166.
41.5	.012	50	27.40	-0.042	44213.
51.0	.010	50	30.86	-0.044	55220.

column gives the corresponding value of p, and the third column the chosen number of colonizing species. The fourth column is the resultant average species richness S of the reshuffled communities (from Section III.4.2), and the last two columns give the computed average values for $\lambda^{(0)}$ and $\lambda^{(1)}$.

These results are briefly discussed in Section III.4.4.

Appendix B.   SOME LOTKA-VOLTERRA FORMULAE

We shall list here some formulae having to do with Lotka-Volterra models

$$\frac{dx_i}{dt} = f_i(x_1,\ldots,x_n) = C_i x_i (k_i - \sum_j \alpha_{ij} x_j) \quad , \quad i=1,\ldots,n$$

for competition among n species. The quantities C_i, k_i, and α_{ij} are positive real constants, with $\alpha_{ii}=1$ for all i.

B.1. One-species equilibria.

We shall first calculate various quantities at an equilibrium with only one species, say species s, present. Then the equilibrium values for the population densities x_i are

(B.1) $x_i = \hat{x}_i(s) = \delta_{is} k_s$,

where δ_{ij} is the Kronecker delta

$$\delta_{ij} = \begin{cases} 1 , & i=j \\ 0 , & i \neq j. \end{cases}$$

The stability matrix A for this equilibrium is easily calculated to be

(B.2) $\quad A_{ij}(s) = \dfrac{\partial f_i}{\partial x_j}(\hat{x}(s)) = \delta_{ij}R_i(s) + \delta_{is}S_j(s)$,

with

(B.3)
$$\begin{cases} R_i(s) = C_i(k_i - \alpha_{is}k_s) \\[2mm] S_j(s) = -C_s\alpha_{sj}k_s . \end{cases}$$

If we make the Ansatz

(B.4) $\quad A_{ij}^{-1}(s) = \delta_{ij}T_i(s) + \delta_{is}U_j(s)$,

we easily find from

$$\sum_j A_{ij}(s)A_{jk}^{-1}(s) = \delta_{ik}$$

that such a solution exists, with

(B.5)
$$\begin{cases} T_i(s) = 1/R_i(s) = 1/C_i(k_i - \alpha_{is}k_s) , \quad i \neq s \\[2mm] U_i(s) = -S_i(s)/R_i(s)S_s(s) = -\alpha_{si}/C_i(k_i - \alpha_{is}k_s) , \quad i \neq s \\[2mm] T_s(s) + U_s(s) = 1/S_s(s) = -1/C_s k_s . \end{cases}$$

Let $v(s)$ be an eigenvector (column matrix) of $A(s)$. Then from equation (2),

$$A(s) \begin{bmatrix} v_1(s) \\ v_2(s) \\ \cdot \\ \cdot \\ \\ \\ \\ \\ \\ \\ \\ \cdot \\ \cdot \\ v_n(s) \end{bmatrix} = \begin{bmatrix} R_1(s)v_1(s) \\ R_2(s)v_2(s) \\ \cdot \\ \cdot \\ \\ \cdot \\ R_{s-1}(s)v_{s-1}(s) \\ \Sigma S_i(s)v_i(s) \\ i \\ R_{s+1}(s)v_{s+1}(s) \\ \cdot \\ \\ \cdot \\ R_n(s)v_n(s) \end{bmatrix} = \lambda(s) \begin{bmatrix} v_1(s) \\ v_2(s) \\ \cdot \\ \cdot \\ \\ \\ \\ \\ \\ \\ \\ \cdot \\ \cdot \\ v_n(s) \end{bmatrix}$$

where $\lambda(s)$ is the eigenvalue corresponding to $v(s)$. It is immediately apparent that n linearly independent solutions to these equations are given by the components

(B.6)
$$(v_i(s))_j = \delta_{ji} + \delta_{js}\beta_i(s) , \quad i \neq s$$

$$(v_s(s))_j = \delta_{js},$$

for some $\beta_i(s)$. Since, then, A has n linearly independent eigenvectors, it is semisimple (Definition A.1). A short calculation yields

(B.7) $\quad \beta_i(s) = S_i(s)/(R_i(s)-S_s(s)) = -C_s\alpha_{si}k_s/(C_sk_s+C_i(k_i-\alpha_{is}k_s))$

(B.8)
$$\begin{cases} \lambda_i(s) = R_i(s) = C_i(k_i-\alpha_{is}k_s) , \quad i \neq s \\ \\ \lambda_s(s) = S_s(s) = -C_sk_s. \end{cases}$$

If $\tilde{v}(s)$ is a left-eigenvector (row matrix) of $A(s)$, then

$$\left[\tilde{v}_1(s) \quad \tilde{v}_2(s) \quad \bullet \quad \bullet \quad \tilde{v}_n(s) \quad A(s)\right] =$$

$$= \left[R_1(s)\tilde{v}_1(s)+S_1(s)\tilde{v}_s(s) \quad \bullet \quad \bullet \quad R_n(s)\tilde{v}_n(s)+S_n(s)\tilde{v}_s(s)\right]$$

$$= \lambda(s)\left[\tilde{v}_1(s) \quad \tilde{v}_2(s) \quad \bullet \quad \bullet \quad \tilde{v}_n(s)\right] .$$

Again, the solutions follow practically from inspection:

$$(B.9) \quad \begin{cases} (\tilde{v}_i(s))_j = \delta_{ji} , \quad i{\neq}s \\[2ex] (\tilde{v}_s(s))_j = S_j(s)/(S_s(s)-R_j(s)) = C_s\alpha_{sj}k_s/(C_sk_s+C_j(k_j-\alpha_{js}k_s)), \end{cases}$$

with the corresponding eigenvalues again given by (8). These eigen-vectors have been normalized so that $\tilde{v}_iv_j=\delta_{ij}$ for all i, j.

Finally, a short calculation shows that

$$(B.10) \quad \frac{\partial^2 f_i}{\partial x_j \partial x_k}(\hat{x}(s)) = -C_i(\delta_{ij}\alpha_{ik}+\delta_{ik}\alpha_{ij}) .$$

B.2. Two-species equilibria.

We shall now consider equilibria with two species, say s and t, present. The equilibrium densities are given by

$$k_s-\hat{x}_s(t)-\alpha_{st}\hat{x}_t(s)=0$$

$$k_t-\alpha_{ts}\hat{x}_s(t)-\hat{x}_t(s)=0 ,$$

where $\hat{x}_s(t)$ is the equilibrium density of species s in equilibrium with species t, and $\hat{x}_t(s)$ is defined analogously. The solutions are

(B.11) $\hat{x}_s(t) = (k_s - \alpha_{st}k_t)/(1-\alpha_{st}\alpha_{ts})$, $\hat{x}_t(s) = (k_t - \alpha_{ts}k_s)/(1-\alpha_{st}\alpha_{ts})$

with, of course, all other species densities zero, so that the n-species equilibrium density is

(B.12) $\hat{x}_i(s,t) = \delta_{is}\hat{x}_s(t) + \delta_{it}\hat{x}_t(s)$.

The stability matrix is

(B.13) $A_{ij}(s,t) = \dfrac{\partial f_i}{\partial x_j}(\hat{x}(s,t)) = \delta_{ij}R_i(s,t) + \delta_{is}S_j(s,t) + \delta_{it}T_j(s,t)$

with

(B.14)
$$\begin{cases} R_i(s,t) = C_i(k_i - \alpha_{is}\hat{x}_s(t) - \alpha_{it}\hat{x}_t(s)) \\[2mm] S_i(s,t) = -C_s\alpha_{si}\hat{x}_s(t) \\[2mm] T_i(s,t) = -C_t\alpha_{ti}\hat{x}_t(s) . \end{cases}$$

Making the Ansatz

(B.15) $A_{ij}^{-1}(s,t) = \delta_{ij}U_i(s,t) + \delta_{is}V_j(s,t) + \delta_{it}W_j(s,t)$,

we find after some algebraic labor

(B.16)
$$\begin{cases} U_i(s,t) = 1/R_i(s,t) , \quad i \neq s,t \\[2mm] U_s(s,t) + V_s(s,t) = T_t(s,t)/(S_s(s,t)T_t(s,t) - S_t(s,t)T_s(s,t)) \\[2mm] U_t(s,t) + W_t(s,t) = S_s(s,t)/(S_s(s,t)T_t(s,t) - S_t(s,t)T_s(s,t)) \end{cases}$$

$$(B.17) \begin{cases} V_i(s,t) = \dfrac{S_t(s,t)T_i(s,t)-S_i(s,t)T_t(s,t)}{R_i(s,t)(S_s(s,t)T_t(s,t)-S_t(s,t)T_s(s,t))} \quad , \quad i \neq s,t \\[18pt] V_t(s,t) = \dfrac{S_s(s,t)}{S_t(s,t)T_s(s,t)-S_s(s,t)T_t(s,t)} \end{cases}$$

$$(B.18) \begin{cases} W_i(s,t) = \dfrac{S_s(s,t)T_i(s,t)-S_i(s,t)T_s(s,t)}{R_i(s,t)(S_t(s,t)T_s(s,t)-S_s(s,t)T_t(s,t))} \quad , \quad i \neq s,t \\[18pt] W_s(s,t) = \dfrac{T_s(s,t)}{S_t(s,t)T_s(s,t)-S_s(s,t)T_t(s,t)} \; . \end{cases}$$

Writing out the equations for the eigenvectors and eigenvalues of $A(s,t)$ explicitly, we find after a little more algebra the eigenvalues

$$(B.19) \begin{cases} \lambda_i(s,t) = R_i(s,t) \quad , \quad i \neq s,t \\[12pt] \lambda_s(s,t) = S_s(s,t)+\gamma_+(s,t)S_t(s,t) \\[12pt] \lambda_t(s,t) = S_s(s,t)+\gamma_-(s,t)S_t(s,t) \end{cases}$$

where

$$(B.20) \quad \begin{aligned} \gamma_\pm(s,t) &= (T_t(s,t)-S_s(s,t)\pm((S_s(s,t)-T_t(s,t))^2+ \\[8pt] &\quad +4T_s(s,t)S_t(s,t))^{1/2})/2S_t(s,t). \end{aligned}$$

The corresponding eigenvectors (column matrices) are

$$(B.21) \begin{cases} (v_i(s,t))_j = \delta_{ji}+\delta_{js}\beta_i(s,t)+\delta_{jt}\varepsilon_i(s,t) \quad , \quad i \neq s,t \\[12pt] (v_s(s,t))_i = (\delta_{is}+\gamma_\pm(s,t)\delta_{it})/(1+S_t(s,t)\gamma_\pm(s,t)^2/T_s(s,t)) \\ t \end{cases}$$

with

$$(B.22) \begin{cases} \beta_i(s,t) = \dfrac{(R_i(s,t)-T_t(s,t))S_i(s,t)+S_t(s,t)T_i(s,t)}{(R_i(s,t)-T_t(s,t))(R_i(s,t)-S_s(s,t))-S_t(s,t)T_s(s,t)} \\[4mm] \epsilon_i(s,t) = \dfrac{(R_i(s,t)-S_s(s,t))T_i(s,t)+S_i(s,t)T_s(s,t)}{(R_i(s,t)-S_s(s,t))(R_i(s,t)-T_t(s,t))-S_t(s,t)T_s(s,t)} \end{cases}$$

and the corresponding left-eigenvectors (row matrices) are

$$(B.23) \begin{cases} (\tilde{v}_i(s,t))_j = \delta_{ji} \quad , \quad i \neq s,t \\[4mm] (\tilde{v}_s(s,t))_i = \dfrac{S_i(s,t)+S_t(s,t)\gamma_\pm(s,t)T_i(s,t)/T_s(s,t)}{S_s(s,t)+\gamma_\pm(s,t)S_t(s,t)-R_i(s,t)} \\ \quad t \end{cases} \quad .$$

These eigenvectors have been normalized so that $\tilde{v}_i v_j = \delta_{ij}$ for all i, j.

The second derivatives of f_i, evaluated at equilibrium, are again given by equation (10).

Appendix C. HARVESTING AND TWO-SPECIES INTERACTIONS

Harvesting influences the dominance controlled communities of
Chapter III mainly by altering the rules for reshuffles in these systems. In order to understand reshuffles, it is necessary to know the
outcome of an attempted invasion of a patch which is predominated by
one species. In the absence of harvesting, this is quite straightforward (see Section III.2). But harvesting changes the topology of the
phase space trajectories, and therefore the outcomes of attempted invasions.

The within-patch dynamics relevant to such invasions is given by
equations of the form

$$\frac{dx_1}{dt} = C_1 x_1 (1-x_1-\alpha_{12}x_2)-H_1(x_1)$$

(C.1)

$$\frac{dx_2}{dt} = C_2 x_2 (1-\alpha_{21}x_1-x_2)-H_2(x_2),$$

where the harvesting rates are of the form (IV.2). We shall study the
systems (1) in this appendix.

If, for instance, species 1 is harvested with harvesting structure S2 (figure IV.2), then the zero x_1-isocline has the form shown in
figure 1, and if it is harvested with structure S3 (figure IV.3), then
this isocline has the form shown in figure 2. The dimensionless parameters ζ_i and κ_i are defined by equations (IV.3).

In addition to the depicted curves for positive densities, the
zero x_1-isocline includes the curve $x_1=0$ (that is, the x_2-axis).
Notice also that dx_1/dt is positive in, and only in, the regions which

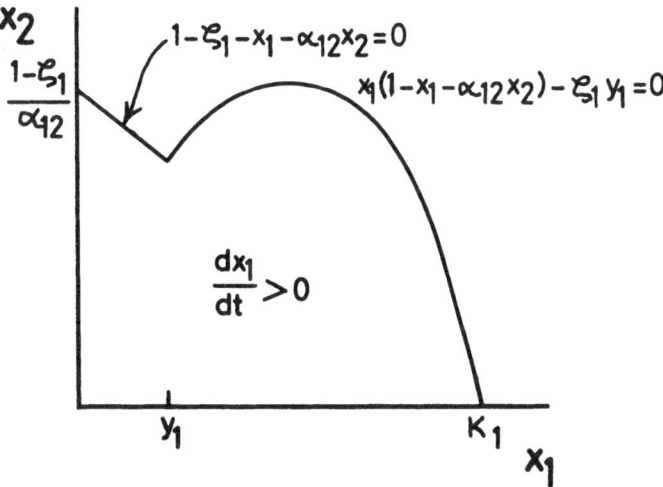

Figure C.1. The zero x_1-isocline for harvesting with structure S2.

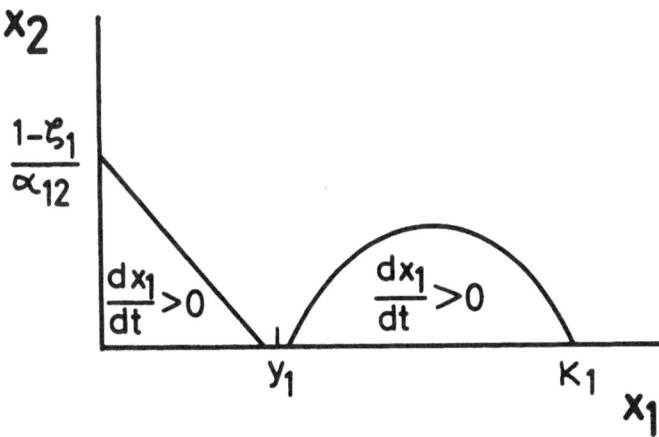

Figure C.2. The zero x_1-isocline for harvesting with structure S3.

are so labelled. (These regions extend onto the adjoining portions of the x_1-axis.)

The zero isoclines for structures S1 and S4 are degenerate forms of these. In the case of structure S1, $y_1 > \kappa_1$ and the zero-isocline consists just of the linear part. In the case of structure S4 the zero isocline has no linear part, but consists just of the parabolic part.

Of course, the x_2-isoclines have similar forms.

*

Now, we want to know what happens if one species (say species 2) invades a patch which is predominated by the other (species 1). In terms of within-patch dynamics, this means that initially the densities (x_1, x_2) are in a neighborhood of the point $(\kappa_1, 0)$ in phase space, with both densities positive, and that the growth rate of species 2, the invading species, is positive at this initial point (for otherwise it cannot invade).

Thus, assume that initially

(C.2) $(x_1, x_2) = (\kappa_1 + \epsilon_1, \epsilon_2)$,

with ϵ_1 and ϵ_2 small and $\epsilon_2 > 0$. Then to first order in ϵ_1, ϵ_2, the growth rate of species 2 is

$$\frac{dx_2}{dt} \simeq C_2 \epsilon_2 (1 - \zeta_2 - \alpha_{21}\kappa_1).$$

Therefore the condition for species 2 to invade is

(C.3) $\kappa_1 < (1 - \zeta_2)/\alpha_{21}$.

Geometrically, this says that the point $(1-\zeta_2)/\alpha_{21}$ on the x_1-axis is to the right of the point κ_1 on that axis. By referring to the above sketches of the isoclines, one easily sees that this indeed places the initial point in the region where $dx_2/dt > 0$.

We shall partition the phase space (excluding the x_1-axis) into five disjoint regions, as follows:

$$R_1 = \{x_1, x_2 \mid y_1 < x_1, \ 0 < x_2 \leqslant y_2\}$$

$$R_2 = \{x_1, x_2 \mid 0 < x_1 \leqslant y_1, \ 0 < x_2 \leqslant y_2\}$$

$$R_3 = \{x_1, x_2 | \ y_1 < x_1, \ y_2 < x_2\}$$

$$R_4 = \{x_1, x_2 | \ 0 < x_1 \leqslant y_1, \ y_2 < x_2\}$$

$$R_5 = \{x_1, x_2 | \ x_1 = 0\}$$

In each of these regions there is at most one stable equilibrium. We then have

Theorem C.1. If there is a stable equilibrium in R_a, $a \in \{1, \ldots, 5\}$, and no stable equilibrium in R_b for all b<a, then the system (1) with the initial values (2) will, for ε_1, ε_2 sufficiently small, tend asymptotically to the stable equilibrium in R_a.

Here "stability" means linearized stability, that is: negativity of the real parts of all eigenvalues of the stability matrix. This is equivalent to asymptotic stability in the sense of Liapunov for all our systems except for a set of measure zero.

The proof is simply to construct all possible phase portaraits (up to homeomorphism) for each of the five cases a=1,...,5, and to verify the claim of the theorem by inspection of these phase portraits. This is a straightforward but lengthy exercise, not all of which will be set down here. Instead, we will first outline some methods which are useful in the proof, and then give explicitly one case (a=2) of the proof.

The linearlized stability matrix has elements

$$A_{ij} = \frac{\partial f_i}{\partial x_j}\bigg|_{x=\hat{x}}$$

for an equilibrium at $x_i = \hat{x}_i$. The characteristic polynomial of this matrix is

$$\lambda^2 - I^0 \lambda + \Omega^0,$$

where

$$I^0 = A_{11} + A_{22} \quad , \quad \Omega^0 = A_{11}A_{22} - A_{12}A_{21} .$$

The condition for linearized stability is that both roots of this polynomial have negative real parts. Necessary and sufficient conditions for this are (Birkhoff and MacLane 1953, Ch. 5, Thm. 10)

$$I^0 < 0 \quad , \quad \Omega^0 > 0 .$$

The condition $\Omega^0 > 0$ has a simple and useful geometric interpretation, which emerges from writing the condition as

$$- \frac{A_{21}}{A_{22}} > - \frac{A_{11}}{A_{12}} .$$

The reader will recognize the left hand side of this inequality as the slope of the zero x_2-isocline at equilibrium, and the right hand side as the slope of the x_1-isocline at equilibrium. Elementary though it is, this observation is worth dignifying as

Lemma C.1. A necessary condition for linearized stability of an equilibrium is that the slope of the zero x_2-isocline is greater than the slope of the zero x_1-isocline at equilibrium. If in addition both of these slopes are negative, this is a sufficient condition for stability.

Proof. We have already proved the first statement. the second statement follows directly from the observation that in the systems treated here, A_{12} and A_{21} are both negative for all positive densities, so that if both slopes are negative then I^0 is negative. ∎

However, these slopes contain still more information. One can read off from them the topology of the system trajectories in a neighborhood of equilibrium:

Lemma C.2. (Sansone and Conti) If the slope of the zero x_2-isocline is greater than that of the zero x_1-isocline at equilibrium

($\Omega^0 > 0$), then the equilibrium is a (stable or unstable) node. If the slope of the zero x_1-isocline is greater ($\Omega^0 < 0$), then the equilibrium is a saddle point (hence, of course, unstable).

   Proof. The proof, in terms of the sign of Ω^0, is given by Sansone and Conti (1964, Ch. 5, Sec. 2); the translation of this into geometric language is simply the above observation about the geometric character of Ω^0. ▮

   These two lemmas are quite useful in the proof of Theorem 1. For example, let us look at

   Proof of Theorem 1, a=2. Our hypothesis is that there is a stable equilibrium in R_2, but none in R_1. In fact, neither can there be a stable equilibrium in R_3, as we now show.

   Suppose for the moment that both species are harvested with harvesting structure S2. Then if there is a stable equilibrium in R_2 we have the following for the linear parts of the isoclines:

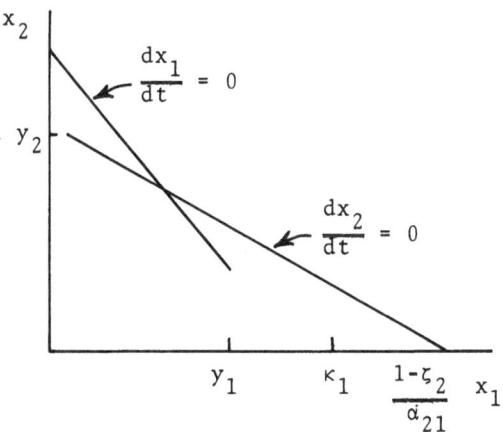

(Use has been made here of Lemma 1.) Now, the x_1-isocline, when extended into the region $x_1 > y_1$, must intersect the x_1-axis at some point $\kappa_1 < (1 - \zeta_2)/\alpha_{21}$: this is the condition for x_2 to invade. But this extended x_1-isocline cannot cross the x_2-isocline in the region $x_1 > y_1$, for this would create a stable equilibrium in R_1 (the stability following from Lemma 1). Hence the x_1-isocline must remain below the x_2-isocline in the region $x_1 > y_1$; in particular, it must remain below $x_2 = y_2$ in the region $x_1 > y_1$. So this isocline never enters R_3 at all, whence there is no equilibrium in R_3.

This leaves us with three structurally stable topologies (one with no equilibrium in R_4, one with one equilibrium in R_4, and one with two equilibria in R_4), and with one possible structurally unstable topology. If there is no equilibrium in R_4, we have

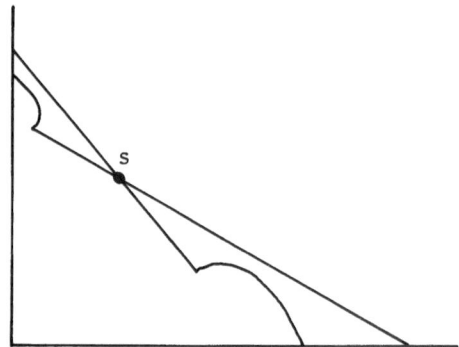

The stable equilibrium s in R_2 is the only stable equilibrium and is globally asymptotically stable (all nontrivial trajectories approach it asymptotically). Hence trajectories with our initial conditions approach s asymptotically.

Moreover, it is clear that none of the trajectories passing through our initial points can pass through R_3, R_4, or R_5 before reaching s. Since the other possible phase space topologies involve changes only in these regions, we see that for all possible topologies the trajectories with our initial data will approach the stable equilibrium in R_2.

It is easily seen that the same argument goes through if one or both species are allowed to have harvesting structure S3; the argument becomes trivial for the degenerate cases S1, S4. ▮

So in order to find the asymptotic behavior of one of these systems, we must search the five regions R_a in turn for stable equilibria. This task can often be simplified by the use of

Theorem C.2. A necessary condition for the occurrence of a stable equilibrium in R_1, R_2, or R_4 is $1 - \alpha_{12}\alpha_{21} > 0$.

Proof. Consider R_1. In that region

$$f_1 = C_1(x_1(1-x_1-\alpha_{12}x_2)-\zeta_1 y_1)$$

$$f_2 = C_2 x_2(1-\zeta_2-\alpha_{21}x_1-x_2)$$

$$-\left[\frac{\partial f_1}{\partial x_1} \bigg/ \frac{\partial f_1}{\partial x_2}\right]\bigg|_{f_1=0} = \frac{\zeta_1 y_1}{\alpha_{12}x_1^2} - \frac{1}{\alpha_{12}}$$

$$-\left[\frac{\partial f_2}{\partial x_1} \bigg/ \frac{\partial f_2}{\partial x_2}\right]\bigg|_{f_2=0} = -\alpha_{21}.$$

From Lemma 1, a necessary condition for stability of any equilibrium in this region is

$$-\alpha_{21} > \frac{\zeta_1 y_1}{\alpha_{12}x_1^2} - \frac{1}{\alpha_{12}},$$

and a necessary condition for this is

$$\frac{1}{\alpha_{12}} > \alpha_{21}$$

$$1-\alpha_{12}\alpha_{21}>0.$$

One proceeds similarly in R_2 and R_4. ∎

Thus, if $1-\alpha_{12}\alpha_{21}<0$, we need only search R_3 and R_5 for stable equilibria.

These results are used in Section IV.2 to formulate the rules for reshuffles of harvested systems.

BIBLIOGRAPHY

Abbott, I., L. K. Abbott, and P. R. Grant. 1977. Comparative ecology of Galápagos ground finches (Geospiza Gould): Evaluation of the importance of floristic diversity and interspecific competition. Ecol. Monogr. 47, 151-184.

Abele, L. G. 1976. Comparative species richness in fluctuating and constant environments: coral-associated decapod crustaceans. Science 192, 461-463.

Adicott, J. F. 1974. Predation and prey community structure: An experimental study of the effect of mosquito larvae on the protozoan communities of pitcher plants. Ecology 55, 475-492.

Ahlfors, L. V. 1953. Complex analysis. McGraw-Hill, New York.

Anderson, D. J. 1961a. The structure of some upland plant communities in Caernarvonshire. I. The pattern shown by Pteridium aquilinum. J. Ecol. 49, 369-377.

Anderson, D. J. 1961b. The structure of some upland plant communities in Caernarvonshire. II. The pattern shown by Vaccinium myrtillus and Calluna vulgaris. J. Ecol. 49, 731-738.

Anderson, R. C., O. L. Loucks, and A. M. Swain. 1969. Herbaceous response to canopy cover, light intensity, and throughfall precipitation in coniferous forests. Ecology 50, 255-263.

Arnold, V. I., and A. Avez. 1970. Ergodic problems in classical mechanics. Benjamin, New York.

Ashby, E. 1948. Statistical ecology, a re-assessment. Bot. Rev. 14, 222-234.

Ashton, P. S. 1969. Speciation among tropical forest trees: some deductions in the light of recent evidence. Biol. J. Linnean Soc. London 1, 155-196.

Aubréville, A. 1938. La forêt coloniale: les forêts de l'Afrique occidentale française. Ann. Acad. Sci. Colon. Paris 9, 1-245.

Auclair, A. N., and F. G. Goff. 1971. Diversity relations of upland forests in the western Great Lakes area. Amer. Natur. 105, 499-528.

Balsler, I., and H. Degn. 1975. Spatial instability in simple reaction schemes. J. Theoret. Biol. 49, 173-177.

Bartholomew, B. 1970. Bare zone between California shrub and grassland communities: the role of animals. Science 170, 1210-1212.

Baumgärtel, H. 1964. Zur Störungstheorie beschränkter linearer Operatoren eines Banachschen Raumes. Math. Nachr. 26, 361-379.

Bazzaz, F. A. 1968. Succession in abandoned fields in the Shawnee Hills, southern Illinois. Ecology 49, 924-936.

Bazzaz, F. A. 1975. Plant species diversity in old-field successional ecosystems in southern Illinois. Ecology 56, 485-488.

Beals, E. W. 1968. Spatial pattern of shrubs on a desert plain in Ethiopia. Ecology 49, 744-746.

Birkhoff, G., and S. MacLane. 1953. A survey of modern algebra. Revised edition. Macmillan, New York.

Bjorkman, O., and P. Holmgren. 1966. Photosynthetic adaptation to light intensity in plants native to shaded and exposed habitats. Physiol. Plantarum 19, 854-859.

Black, G. A., T. Dobzhansky, and C. Pavan. 1950. Some attempts to estimate species diversity and population density of trees in Amazonian forests. Bot. Gaz. 111, 413-425.

Blondel, J., C. Ferry, and B. Frochot. 1973. Avifaune et végétation. Essai d'analyse de la diversité. Alauda. 41, 63-84.

Bochner, S., and W. T. Martin. 1948. Several complex variables. Princeton University Press, Princeton.

Bratton, S. P. 1976. Resource division in an understory herb community: responses to temporal and microtopographic gradients. Amer. Natur. 110, 679-693.

Braun-Blanquet, J. 1932. Plant sociology: the study of plant communities. Translated, revised, and edited by G. D. Fuller and H. S. Conrad. McGraw-Hill, New York.

Brereton, A. J. 1971. The structure of the species populations in the initial stages of salt-marsh succession. J. Ecol. 59, 321-338.

Brooks, J. L., and S. I. Dodson. 1965. Predation, body size and composition of plankton. Science 150, 28-35.

Brues, C. T. 1946. Insect dietary. Harvard University Press, Cambridge, Massachusetts.

Byer, M. D. 1970. Interspecific plant association as influenced by three variables. Ecology 51, 103-112.

Cable, D. R. 1969. Competition in the semidesert grass-shrub type as influenced by root systems, growth habits, and soil moisture extraction. Ecology 50, 27-38.

Cain, S. A., G. M. De Oliviera Castro, J. Murca Piras, and N. T. da Silve. 1956. Application of some phytosociological techniques to Brazilian rain forest. Amer. J. Bot. 43, 911-941.

Cairns, J., K. L. Dickson, and W. H. Yongue. 1971. The consequences of non-selective periodic removal of portions of fresh-water protozoan communities. Trans. Amer. Micros. Soc. 90, 71-80.

Casten, R. G., and T. J. Case. 1978. The effect of dispersal on the stability of some model ecological communities. MS.

Caswell, H. 1978. Predator-mediated coexistence: a non-equilibrium model. Amer. Natur. 112, 127-154.

Clements, F. E. 1916. Plant succession: an analysis of the development of vegetation. Carnegie Inst. Wash. Publ. 242.

Coddington, E. A., and N. Levinson. 1955. Theory of ordinary differential equations. McGraw-Hill, New York.

Cody, M. L. 1970. Chilean bird distribution. Ecology 51, 455-463.

Cody, M. L. 1974. Optimization in ecology. Science 183, 1156-1164.

Coe, M. J., D. H. Cumming, and J. Phillipson. 1976. Biomass and production of large African herbivores in relation to rainfall and primary production. Oecologia 22, 341-354.

Cohen, J. E. 1970. A Markov contingency table model for replicated Lotka-Volterra systems near equilibrium. Amer. Natur. 104, 547-559.

Connell, J. H. 1961a. The influence of interspecific competition and other factors on the distribution of the barnacle Chthamlus stellatus. Ecology 42, 710-723.

Connell, J. H. 1961b. Effects of competition, predation by Thais lapillus, and other factors on natural populations of the barnacle Balanus balanoides. Ecol. Monogr. 31, 61-104.

Connell, J. H. 1970a. A predator-prey system in the marine intertidal region. I. Balanus glandula and several predatory species of Thais. Ecol. Monogr. 40, 49-78.

Connell, J. H. 1970b. On the role of natural enemies in preventing competitive exclusion in some marine animals and in rain forest trees. Proc. Adv. Study Inst. Dynamics Numbers Popul. (Oosterbeek, 1970), 298-312.

Connell, J. H. 1978. Diversity in tropical rain forests and coral reefs. Science 199, 1302-1310.

Conway, E., D. Hoff, and J. Smoller. 1977. Large time behavior of solutions of systems of nonlinear reaction-diffusion equations. MS.

Conway, E., and J. Smoller. 1977. Diffusion and the classical ecological interactions: asymptotics. MS.

Cramer, N. F., and R. M. May. 1972. Interspecific competition, predation and species diversity: a comment. J. Theor. Biol. 34, 289-293.

Darwin, C. R. 1859. On the origin of species by natural selection. John Murray, London.

Day, R. J. 1972. Stand structure, succession, and the use of southern Alberta's Rocky Mountain forest. Ecology 53, 472-478.

Dayton, P. K. 1971. Competition, disturbance, and community organization: the provision and subsequent utilization of space in a rocky intertidal community. Ecol. Monogr. 41, 351-389.

Dayton, P. K. 1972. Toward an understanding of community resilience and the potential effects of enrichments to the benthos at McMurdo Sound, Antarctica. in B. C. Parker (ed.), Proc. Colloq. Conserv. Probl. Antarctic. Allen Press, Lawrence, Kansas.

Dayton, P. K. 1973a. Dispersion, dispersal, and persistence of the annual intertidal alga, Postelsia palmaeformis Ruprecht. Ecology 54, 433-438.

Dayton, P. K. 1973b. Two cases of resource partitioning in an intertidal community: making the right prediction for the wrong reason. Amer. Natur. 107, 662-670.

Dayton, P. K. 1975a. Experimental studies of algal canopy interactions in a sea otter-dominated kelp community at Amchitka Island, Alaska. Fishery Bull. 73, 230-237.

Dayton, P. K. 1975b. Experimental evaluation of ecological dominance in a rocky intertidal algal community. Ecol. Monogr. 45, 137-159.

Diamond. J. M. 1975. Assembly of species communities. pp. 342-444 in Ecology and evolution of communities, M. L. Cody and J. M. Diamond (eds.), Harvard University Press, Cambridge, Massachusetts.

Dieudonné, J. 1969. Foundations of modern analysis. Academic Press, New York.

Dittmer, H. J. 1937. A quantitative study of the roots and root hairs of a winter rye plant (Secale cereale). Amer. J. Bot. 24, 417-420.

Dittmer, H. J. 1938. A quantitative study of the subterranean members of three field grasses. Amer. J. Bot. 25, 654-657.

Dobzhansky, T. 1950. Evolution in the tropics. Amer. Sci. 38, 209-221.

Drew, W. B. 1942. The vegetation of abandoned cropland in Cedar Creek area, Boone and Calaway Counties, Missouri. Univ. Missouri Coll. Agric. Exp. Stn. Research Bull. 344.

Drury, W. H., and I. C. T. Nisbet. 1973. Succession. J. Arnold Arboretum Harvard University 54, 331-368.

Du Rietz, G. E. 1930. Classification and nomenclature of vegetation. Svensk botanisk tidskrift 24, 489-503.

Eggeling, W. J. 1947. Observations on the ecology of Budongo rain forest, Uganda. J. Ecol. 34, 20-87.

Eggers, D. M. 1977. The nature of prey selection by planktivorous fish. Ecology 58, 46-59.

Ehrlich, P. R., and L. C. Birch. 1967. The "Balance of Nature" and "population control". Amer. Natur. 101, 97-107.

Elton, C. S. 1927. Animal ecology. Macmillan, New York.

Emlen, J. M. 1966. The role of time and energy in food preference. Amer. Natur. 100, 611-617.

Emlen, J. M. 1968. Optimal choice in animals. Amer. Natur. 102, 385-389.

Emlen, J. M. 1973. Ecology: an evolutionary approach. Addison-Wesley, Reading.

Evans, R. A. 1960. Differential responses of three species of the annual grassland type to plant competition and mineral nutrition. Ecology 41, 305-310.

Fisher-Piette, E. 1935. Histoire d'une moulière. Bull. Biol. 69, 154-180.

Fox, J. E. D. 1970. Natural regeneration of the Kambui Hills forest in eastern Sierra Leone. I. Ecological status of Lophira/Heritiera rain forest. Trop. Ecol. 11, 169-185.

Fox, J. F. 1977. Alternation and the coexistence of tree species. Amer. Natur. 111, 69-89.

Fox, L. 1964. An introduction to numerical linear algebra. Oxford University press, Oxford.

Gant, R. E. and E. E. C. Clebsch. 1975. The allelopathic influences of Sassafras albidum in old-field succession in Tenessee. Ecology 56, 604-615.

Gause, G. F. 1934. The struggle for existence. Williams and Wilkins, Baltimore.

Gilpin, M. E. 1975. Limit cycles in competition communities. Amer. Natur. 109, 51-60.

Gilpin, M. E., and T. J. Case. 1976. Multiple domains of attraction in competition communities. Nature 261, 40-42.

Gillet, J. B. 1962. Pest pressure, an underestimated factor in evolution. Systematics Association Publ. No. 4, 37-46.

Goodall, D. W. 1970. Statistical plant ecology. Ann. Rev. Ecol. Syst. 1, 99-124.

Glynn, P. W. 1973. Acanthaster: effect on coral reef growth in Panama. Science 180, 504-506.

Glynn, P. W. 1976. Some physical and biological determinants of coral community structure in the eastern Pacific. Ecol. Monogr. 46, 431-456.

Green, T. W., and I. G. Palmbad. 1975. Effects of insect seed predators on Astragalus cibarius and Astragalus utahensis (Leguminosae).
Ecology 56, 1435-1440.

Grieg-Smith, P. 1961. Data on pattern within plant communities. II.
Ammophila arenaria (L.) Link. J. Ecol. 49, 703-708.

Grigg, R. W., and J. R. Maragos. 1974. Recolonization of hermatypic
corals on submerged lava flows in Hawaii. Ecology 55, 387-395.

Grime, J. P. 1977. Evidence for the existence of three primary strategies in plants and its relevance to ecological and evolutionary
theory. Amer. Natur. 111, 1169-1194.

Habeck, J. R., and R. W. Mutch. 1973. Fire dependent forests in the
northern Rocky Mountains. Quat. Res. 3, 408-424.

Hahn, W. 1967. Stability of motion. Springer-Verlag, Berlin.

Hairston, N. G. 1959. Species abundance and community organization.
Ecology 40, 404-416.

Hairston, N. G., F. Smith, and L. B. Slobodkin. 1960. Community structure, population control, and competition. Amer. Natur. 94, 421-
425.

Halmos, P. R. 1958. Finite-dimensional vector spaces. Van Nostrand,
Princeton.

Hall, D. J., W. E. Cooper, and E. E. Werner. 1970. An experimental
approach to the production dynamics and structure of freshwater
animal communities. Limnol. Oceanogr. 15, 839-928.

Hall, J. B. 1971. Pattern in a chalk grassland community. J. Ecol.
59, 749-762.

Harner, R. F., and K. T. Harper. 1976. The role of area, heterogeneity
and favorability in plant species diversity of Pinyon-Juniper ecosystems. Ecology 57, 1254-1263.

Harper, J. L. 1969. The role of predation in vegetational diversity.
Brookhaven Symposium in Biology No. 22, Diversity and Stability in
Ecological Systems, 48-62.

Harper, J. L. 1977. Population biology of plants. Academic Press,
London.

Harper, J. L., J. T. Williams, and G. R. Sagar. 1965. The behaviour
of seeds in soil. J. Ecol. 51, 273-286.

Harris, G. A., and A. M. Wilson. 1970. Competition for moisture among
seedlings of annual and perennial grasses as influenced by root
elongation at low temperature. Ecology 51, 530-534.

Hart, D. D. 1977. The response of stream benthic invertebrate communities to microspatial substrate complexity. Bull. Ecol. Soc.
Amer. 58, 37.

Heilman, P. E. 1966. Change in distribution and availability of nitrogen with forest succession on north slopes in interior Alaska.
Ecology 47, 825-831.

Heilman, P. E. 1968. Relationship of availability of phosphorous and
cations to forest succession and bog formation in interior Alaska.
Ecology 49, 331-336.

Heinselman, M. L. 1973. Fire in the virgin forests of the Boundary
Waters Canoe Area, Minnesota. Quat. Res. 3, 329-382.

Hewetson, C. E. 1956. A discussion on the climax concept in relation
to the tropical rain and deciduous forest. Emp. For. Rev. 35, 274-
291.

Hlavácek, V., J. Sinkúle, and M. Kubícek. 1972. A method for determining effects of diffusion on auto-catalytic chemical reactions.
J. Theoret. Biol. 36, 283-290.

Holling, C. S. 1965. The functional response of predators to prey
density and its role in mimicry and population regulation. Mem.
Ent. Soc. Canada 45, 5-60.

Holling, C. S. 1973. Resilience and stability of ecological systems.
Ann. Rev. Ecol. Syst. 4, 1-23.

Hope-Simpson, J. F. 1940. Studies of the vegetation of the English
chalk. VI. Late stages in succession leading to chalk grassland.
J. Ecol. 28, 386-402.

Hopkins, B. 1955. The species area relations of plant communities.
J. Ecol. 43, 409-426.

Horn, H. S. 1971. The adaptive geometry of trees. Princeton University press, Princeton.

Horn, H. S. 1974. The ecology of secondary succession. Ann. Rev.
Ecol. Syst. 5, 25-37.

Horn, H. S. 1975. Markovian processes of forest succession. in
Ecology and evolution of communities, M. L. Cody and J. M. Diamond
(eds.), Harvard University Press, Cambridge, Massachusetts.

Horn, H. S. 1976. Succession. in Theoretical ecology, R. M. May (ed.)
Blackwell Scientific Publications, Oxford.

Horn, H. S., and R. H. MacArthur. 1972. Competition among fugitive
species in a harlequin environment. Ecology 53, 749-752.

Hutchinson, G. E. 1951. Copepodology for the ornithologist. Ecology
32, 571-577.

Hutchinson, G. E. 1959. Homage to Santa Rosalia, or Why are there so
many kinds of animals? Amer. Natur. 93, 145-159.

Ivlev, V. S. 1961. Experimental ecology of the feeding of fishes.
Yale University Press, New Haven.

Jackson, J. B. C. 1977a. Competition on marine hard substrata: the adaptive significance of solitary and colonial strategies. Amer. Natur. 111, 743-767.

Jackson, J. B. C. 1977b. Habitat area, Colonization, and development of epibenthic community structure. Pages 349-358 in Proceedings of the 11th European marine biology symposium. Pergamon, Oxford.

Jackson, J. B. C., and L. Buss. 1975. Allelopathy and spatial competition among coral reef invertebrates. Proc. Nat. Acad. Sci. USA 72, 5160-5163.

Janzen, D. H. 1966. Coevolution of mutualism between ants and acacias in Central America. Evolution 20, 249-275.

Janzen, D. H. 1967. Interaction of the bull's-horn acacia (Acacia cornigera L.) with an ant inhabitant (Pseudomyrmex ferruginea F. Smith) in eastern Mexico. Univ. Kansas Sci. Bull. 47, 315-558.

Janzen, D. H. 1969. Allelopathy by myrmecophytes: the ant Azteca as an allelopathic agent of Cecropia. Ecology 50, 147-153.

Janzen, D. H. 1970. Herbivores and the number of tree species in tropical forests. Amer. Natur. 104, 501-528.

Janzen, D. H. 1972. Escape in space by Sterculia apetala seeds from the bug Dysdercus fasciatus in a Costa Rican deciduous forest. Ecology 53, 350-361.

Johnson, N. L., and S. Kotz. 1969. Discrete distributions. Houghton Mifflin Company, Boston.

Jones, N. S. 1948. Observations and experiments on the biology of Patella vulgata at Port St. Mary, Isle of Man. Proc. Trans. Liverpool Biol. Soc. 56, 60-77.

Kain, J. M., and N. S. Jones. 1966. Algal colonization after removal of Echinus. Proceedings of the fifth international seaweed symposium, Halifax, Canada. Pergamon Press, Oxford. pp. 139-140.

Kan-Ichi Sakai. 1961. Competitive ability in plants: its inheritance and some related problems. Symp. Soc. Exp. Biol. 15, 245-263.

Karr, J. R. 1971. Structure of avian communities in selected Panama and Illinois habitats. Ecol. Monogr. 41, 207-233.

Karr, J. R., and R. R. Roth. 1971. Vegetation structure and avian diversity in several new world areas. Amer. Natur. 105, 423-435.

Kato, T. 1966. Perturbation theory for linear operators. Springer-Verlag, Berlin.

Keever, C. 1950. Causes of succession on old fields of the Piedmont, North Carolina. Ecol. Monogr. 20, 230-250.

Kershaw, K. A. 1958. An investigation of the structure of a grassland community. I. The pattern of Agrostis tenuis. J. Ecol. 46, 571-592.

Kershaw, K. A. 1962. Quantitative ecological studies from Landmanna-hellir, Iceland. III. The variation of performance of Carex bige-lowii. J. Ecol. 50, 393-399.

Kershaw, K. A. 1963. Pattern in vegetation and its causality. Ecology 44, 493-499.

Kershaw, K. A. 1973. Quantitative and dynamic plant ecology. Second edition. Edward Arnold, London.

Knight, D. H. A phytosociological analysis of species-rich tropical forest on Barro Colorado Island, Panama. Ecol. Monogr. 45, 259-284.

Kohn, A. J., and P. J. Leviten. 1976. Effect of habitat complexity on population density and species richness in tropical intertidal gas-tropod assemblages. Oecologia 25, 199-210.

Kolmogoroff, A., I. Petrovskij, and N. Piskunov. 1937. Etude de l'équation de la diffusion avec croissance de la quantité de ma-tière et son application à une problème biologique. Bull. Univ. Moscow, Ser. Internat., Ser. A, 6, 1-25.

Krishnamurthy, L. 1977. Functional properties and organization of grazing lands ecosystem. J. Theoret. Biol. 68, 65-72.

Landenberger, D. E. 1968. Studies on selective feeding in the Pacific starfish, Pisaster, in southern California. Ecology 49, 1062-1075.

Lang, J. C. 1972. Interspecific aggression by scleractinian corals. I. The rediscovery of Scolymia cubensis (Milne Edwards and Haime). Bull. Marine Sci. 21, 952-959.

Lang, J. C. 1973. Interspecific aggression by scleractinian corals. II. Why the race is not only to the swift. Bull. Marine Sci. 23, 260-279.

LaSalle, J. P., and S. Lefschetz. 1961. Stability by Liapunov's direct method. Academic Press, New York.

Lazenby, A. 1955. Germination and establishment of Juncus effusus. I. The effect of different companion species and of variation in soil and fertility conditions. J. Ecol. 43, 103-119.

Leighton, D. H., L. G. Jones, and W. J. North. 1966. Ecological rela-tionships between giant kelp and sea urchins in southern California. Proceedings of the fifth international seaweed symposium, Halifax, Canada. Pergamon Press, Oxford. pp. 141-153.

Levin, S. A. 1974. Dispersion and population interactions. Amer. Natur. 108, 207-228.

Levin, S. A. 1976a. Spatial patterning and the structure of ecological communities. in Some mathematical questions in biology, S. A. Levin (ed.), 7, 1-36. Lectures on mathematics in the life sciences, Vol. 8. Amer. Math. Soc., Providence, Rhode Island.

Levin, S. A. 1976b. Population dynamic models in heterogeneous envi-ronments. Ann. Rev. Ecol. Syst. 7, 287-310.

Levin, S. A. 1978a. Population models and community structure in heterogeneous environments. To appear in Mathematical Association of America Studies in Mathematical Biology, Volume II: Populations and communities (S. A. Levin, ed.).

Levin, S. A. 1978b. On the evolution of ecological parameters. in P. F. Brussard (ed.), Proceedings of the symposium on genetics and ecology: the interface. Springer-Verlag, Berlin.

Levin, S. A., and R. T. Paine. 1974. Disturbance, patch formation, and community structure. Proc. Nat. Acad. Sci. USA 71, 2744-2747.

Levin, S. A., and R. T. Paine. 1975. The role of disturbance in models of community structure. in S. A. Levin (ed.), Ecosystem analysis and prediction, Society for Industrial and Applied Mathematics, Philadelphia.

Levins, R., and D. Culver. 1971. Regional coexistence of species and competition between rare species. Proc. Nat. Acad. Sci. USA 68, 1246-1248.

Lieth, H. 1960. Patterns of change within grassland communities. Pages 27-39 in J. L. Harper (ed.), British Ecological Society Symposium No. 1. The biology of weeds. Blackwell's, Oxford.

Lindeman, R. L. 1942. The trophic dynamic aspect of ecology. Ecology 23, 399-418.

Lloyd, M. 1967. Mean crowding. J. Anim. Ecol. 36, 1-30.

Lloyd, M., R. F. Inger, and F. W. King. 1968. On the diversity of reptile and amphibean species in a Bornean rain forest. Amer. Natur. 102, 497-515.

Loucks, O. L. 1970. Evolution of diversity, efficiency, and community stability. Amer. Zool. 10, 17-25.

Lubchenco Menge, J. 1978. Plant species diversity in a marine intertidal community: importance of herbivore food preference and algal competitive abilities. Amer. Natur. 112, 23-39.

Luckens, P. A. 1970. Breeding, settlement and survival of barnacles at artificially modified shore levels at Leigh, New Zealand. New Zealand J. Mar. Freshwater Res. 4, 497-514.

MacArthur, R. H. 1972. Geographical ecology. Harper and Row, New York

MacArthur, R. H., and J. W. MacArthur. 1961. On bird species diversity. Ecology 42, 594-598.

MacArthur, R. H., H. Recher, and M. L. Cody. 1966. On the relation between habitat selection and species diversity. Amer. Natur. 100, 319-332.

Maguire, L. A., and J. W. Porter. 1977. A spatial model of growth and competition strategies in coral communities. Ecol. Modelling 3, 249-271.

May, R. M. 1973. Stability in randomly fluctuating versus deterministic environments. Amer. Natur. 107, 621-650.

May, R. M. 1974. Stability and complexity in model ecosystems. Second edition. Princeton University Press, Princeton.

May, R. M. 1975. Patterns of species abundance and diversity. in M. L. Cody and J. M. Diamond (eds.), Ecology and evolution of communities. Harvard University Press, Cambridge.

May, R. M. 1976. Patterns in multi-species communities. in Theoretical ecology, R. M. May (ed.), Blackwell, Oxford.

May, R. M. 1977. Thresholds and breakpoints in ecosystems with a multiplicity of stable states. Nature 269, 471-477.

May, R. M., and W. J. Leonard. 1975. Nonlinear aspects of competition between three species. SIAM J. Appl. Math. 29, 243-253.

May, R. M., and G. F. Oster. 1976. Bifurcations and dynamic complexity in simple ecological models. Amer. Natur. 110, 573-599.

Maynard Smith, J. 1974. Models in Ecology. Cambridge University Press, Cambridge.

McCowan, R. L., and W. A. Williams. 1968. Competition for nutrients and light between the annual grassland species Bromus mollis and Erodium botrys. Ecology 49, 981-990.

Menge, B. A. 1976. Organization of the New England rocky intertidal community: role of predation, competition, and environmental heterogeneity. Ecol. Monogr. 46, 355-393.

Monk, C. D. 1967. Tree species diversity in the eastern deciduous forest with particular reference to north central Florida. Amer. Natur. 101, 173-187.

Mooney, H. A., and E. L. Dunn. 1970. Photosynthetic systems of Mediterranean climate shrubs and trees of California and Chile. Amer. Natur. 104, 447-453.

Moore, C. W. E. 1959. The competitive effect of Danthonia spp. on the establishment of Bothriochloa ambigua. Ecology 40, 141-143.

Motomura, I. 1932. A statistical treatment of association. (in Japanese) Japan. J. Zool. 44, 379-383.

Muller, C. H. 1966. The role of chemical inhibition (allelopathy) in vegetational composition. Bull. Torrey Bot. Club 93, 332-351.

Muller, C. H., and R. del Moral. 1971. Role of animals in suppression of herbs by shrubs. Science 173, 462-463.

Murdoch, W. W. 1966. "Community structure, population control, and competition" — a critique. Amer. Natur. 100, 219-226.

Murdoch, W. W. 1969. Switching in general predators: experiments on predator specificity and stability of prey populations. Ecol. Monogr. 39, 335-354.

Murdoch, W. W., F. C. Evans, and C. H. Peterson. 1972. Diversity and pattern in plants and insects. Ecology 53, 819-829.

Nagel, J.-L. 1950. Changement d'essences. J. For. Suisse 101, 95-104.

Nash, T. H. III. 1975. Influence of effluents from a zinc factory on lichens. Ecol. Monogr. 45, 183-198.

Nicholson, S. A., and C. D. Monk. 1974. Plant species diversity in old-field succession on the Georgia Piedmont. Ecology 55, 1075-1085.

Nicholson, S. A., and J. T. Scott. 1972. Successional trends in plant communities in the Lake George drainage basin. Bull. Ecol. Soc. Amer. 53, 17.

Nicolis, G., and I. Prigogine. 1977. Self-organization in nonequilibrium systems. Wiley, New York.

Nicotri, M. E. 1977. Grazing effects of four marine intertidal herbivores on the microflora. Ecology 58, 1020-1032.

Odum, E. P. 1960. Organic production and turnover in old-field succession. Ecology 41, 34-49.

Odum, E. P. 1963. Ecology. Holt, Rinehart, and Winston, New York.

Oosting, H. J. 1942. An ecological analysis of the plant communities of Piedmont, North Carolina. Am. Midl. Nat. 28, 1-126.

Osman, R. W. 1977. The establishment and development of a marine epifaunal community. Ecol. Monogr. 47, 37-63.

Othmer, H. G., and L. E. Scriven. 1971. Instability and dynamic pattern in cellular networks. J. Theoret. Biol. 32, 507-537.

Overland, L. 1966. The role of allelopathic substances in the "smother crop" barley. Am. J. Bot. 53, 423-432.

Paine, R. T. 1966. Food web complexity and species diversity. Amer. Natur. 100, 65-75.

Paine, R. T. 1971. A short-term experimental investigation of resource partitioning in a New Zealand rocky intertidal habitat. Ecology 52, 1096-1106.

Paine, R. T. 1974. Intertidal community structure. Experimental studies on the relationship between a dominant competitor and its principal predator. Oecologia 15, 93-120.

Paine, R. T., and R. L. Vadas. 1969. The effects of grazing by sea urchins, Strongylocentrotus spp., on benthic algal populations. Limnol. Oceanogr. 14, 710-719.

Paijmans, K. 1970. An analysis of four tropical rain forest sites in New Guinea. J. Ecol. 58, 77-101.

Parrish, J. A. D., and F. A. Bazzaz. 1976. Underground niche separation in successional plants. Ecology 57, 1281-1288.

Parrish, J. D., and S. B. Saila. 1970. Interspecific competition and species diversity. J. Theoret. Biol. 27, 207-220.

Pearson, G. A. 1942. Herbaceous vegetation, a factor in natural regeneration of Ponderosa Pine in the south-west. Ecol. Monogr. 12, 315-338.

Peterkin, G. F., and C. R. Tubbs. 1965. Woodland regeneration in the New Forest, Hampshire, since 1650. J. Appl. Ecol. 2, 159-170.

Petrusewicz, K., and A. Macfadyen. 1970. Productivity of terrestrial animals: principles and methods. IBP Handbook 13. Davis, Philadelphia.

Phillipson, J. 1975. Rainfall, primary production, and "carrying capacity" of Tsavo National Park, Kenya. E. Afr. Wildlife J. 13, 171-201.

Pianka, E. R. 1966. Latitudinal gradients in species diversity: a review of concepts. Amer. Natur. 100, 33-46.

Pianka, E. R. 1967. Lizard species diversity. Ecology 48, 333-351.

Pianka, E. R. 1976. Competition and niche theory. in R. M. May (ed.), Theoretical ecology, Blackwell, Oxford.

Pianka, E. R., and R. B. Huey. 1971. Bird species diversity in the Kalahari and the Australian deserts. Koedoe 14, 123-130.

Pielou, E. C. 1969. An introduction to mathematical ecology. Wiley, New York.

Pielou, E. C. 1975. Ecological diversity. Wiley, New York.

Pires, J. M., T. Dobzhansky, and G. A. Black. An estimate of the number of species of trees in an Amazonian forest community. Bot. Gaz. 114, 467-477.

Platt, W. J., and I. M. Weis. 1977. Resource partitioning within a guild of fugitive prairie plants. Amer. Natur. 111, 749-513.

Poore, M. E. D. 1964. Integration in the plant community. J. Anim. Ecol. 33 (suppl.), 213-226.

Poore, M. E. D. 1968. Studies in Malaysian rain forest. I. The forest on triassic sediments in Jengka Forest Reserve. J. Ecol. 56, 143-196.

Porter, J. W. 1972. Predation by Acanthaster and its effect on coral species diversity. Amer. Natur. 106, 487-492.

Porter, J. W. 1974. Community structure of coral reefs on opposite sides of the Isthmus of Panama. Science 186, 543-545.

Porter, J. W. 1976. Autotrophy, heterotrophy, and resource partitioning in Caribbean reef-building corals. Amer. Natur. 110, 731-742.

Potzger, J. E., and E. Potzger. 1950. Secondary succession in an Illinoian tillplain habitat. Proc. Indiana Acad. Sci. 59, 95-101.

Quarterman, E. 1957. Early plant succession on abandoned cropland in the central basin of Tennessee. Ecology 38, 300-309.

Randall, J. E. 1965. Grazing effect on sea grasses by herbivorous reef fishes in the West Indies. Ecology 46, 255-260.

Recher, H. F. 1969. Bird species diversity and habitat diversity in Australia and North America. Amer. Natur. 103, 75-80.

Reiners, W. A., I. A. Worley, and D. B. Lawrence. 1971. Plant diversity in a chronosequence at Glacier Bay, Alaska. Ecology 52, 55-69.

Rice, E. L. 1968. Inhibition of nodulation on inoculated legumes by pioneer plant species from abandoned fields. Bull. Torrey Bot. Club 95, 346-358.

Richards, P., and G. B. Williamson. 1975. Treefalls and patterns of understory species in a wet lowland tropical forest. Ecology 56, 1226-1229.

Richards, P. W. 1952. The tropical rainforest. Cambridge University Press, Cambridge.

Risser, P. G., and E. L. Rice. 1971. Diversity in tree species in Oklahoma upland forests. Ecology 52, 876-880.

Rollet, B. 1969. La régéneration naturelle en forêt dense humide sempervirente de plaine de la Guyane Venezuelienne. Bois. For. Trop. 124, 19-38.

Roughgarden, J. 1974. Species packing and the competition function with illustrations from coral reef fish. Theoret. Pop. Biol. 5, 163-186.

Roughgarden, J., and M. Feldman. 1975. Species packing and predation pressure. Ecology 56, 489-492.

Rosenzweig, M. L., and J. Winakur. 1966. Population ecology of desert rodent communities: habitats and environmental complexity. Ecology 50, 558-572.

Ruelle, D., and F. Takens. 1971. On the nature of turbulence. Comm. Math. Phys. 20, 167-192.

Rützler, K. 1965. Systematik und Oekologie der Poriferen aus Litoral-Schattengebieten der Nordadria. Z. Morph. Oekol. Tiere 55, 1-82.

Sarà M. 1970. Competition and cooperation in sponge populations. Symp. Zool. Soc. London 25, 273-284.

Sale, P. F. 1972. Influence of corals in the dispersion of the pomacentrid fish Dascyllus aruanus. Ecology 53, 741-744.

Sale, P. F. 1975. Patterns of use of space in a guild of territorial reef fishes. Marine Biol. 29, 89-98.

Sale, P. F. 1977. Maintenance of high diversity in coral reef fish communities. Amer. Natur. 111, 337-359.

Sale, P. F., D. C. Potts, and E. Frankel. 1976. Recent studies on
Acanthaster planci. Search 7, 334-338.

Sansone, G., and R. Conti. 1964. Non-linear differential equations.
Pergamon Press, London.

Schaeffer, R., and R. Moreau. 1958. L'alternance des essences. Soc.
For. France-Compté, Bull. 29, 1-12, 76-84, 277-298.

Schjelderup-Ebbe, T. 1935. Social behavior of birds. in C. Murchison
(ed.), A handbook of social psychology. Clark University Press,
Worchester.

Schulz, J. P. 1960. Ecological studies on rain forest in northern
Suriname. Verh. K. Ned. Akad. Wet. 53, 1-367.

Seigler, D., and P. W. Price. 1976. Secondary compounds in plants:
primary functions. Amer. Natur. 110, 101-105.

Shafi, M. I., and G. A. Yarranton. 1973. Diversity, floristic richness, and species evenness during a secondary (post-fire) succession. Ecology 54, 897-902.

Simpson, E. H. 1949. Measurement of diversity. Nature 163, 688.

Sims, H. P., and D. Mueller-Dombois. 1968. Effect of grass competition
and depth to water table on height growth of coniferous tree seedlings. Ecology 49, 597-603.

Sinclair, A. R. E. 1974. Natural regulation of a buffalo population
in east Africa. E. Afr. Wildlife J. 12, 135-154.

Singh, J. S., and R. Misra. 1969. Diversity, dominance, stability,
and net production in the grasslands at Varanasi, India. Can. J.
Bot. 47, 425-427.

Skellam, J. G. 1951. Random dispersal in theoretical populations.
Biometrika 38, 196-218.

Slatkin, M. 1974. Competition and regional coexistence. Ecology 55,
128-134.

Slobodkin, L. B. 1961. Growth and regulation of animal populations.
Holt, Rinehart, and Winston, New York.

Slobodkin, L. B. 1964. Experimental populations of Hydrida. J. Anim.
Ecol. 33 (suppl.), 131-148.

Smith, A. P. 1973. Stratification of temperate and tropical forests.
Amer. Natur. 107, 671-683.

Smith, C. L., and J. C. Tyler. 1972. Space resource sharing in a
coral reef fish community. Bull. Nat. Hist. Mus. Los Angeles City
14, 125-170.

Sparling, J. H. 1967. Assimilation rates of some woodland herbs in
Ontario. Bot. Gaz. 128, 160-168.

Summerhayes, V. S. 1941. The effect of voles (<u>Microtus</u> <u>agrestis</u>) on vegetation. J. Ecol. 29, 14-48.

Sutherland, J. P. 1974. Multiple stable points in natural communities. Amer. Natur. 108, 859-873.

Tansley, A. G., and R. S. Adamson. 1925. Studies of the vegetation of the English chalk. III. The chalk grasslands of the Hampshire-Sussex border. J. Ecol. 13, 177-223.

Taylor, D. L. 1973. Some ecological implications of forest fire control in Yellowstone National Park, Wyoming. Ecology 54, 1394-1396.

Terborgh, J. 1973. On the notion of favorableness in plant ecology. Amer. Natur. 107, 481-501.

Thomas, A. S. 1960. Changes in vegetation since the advent of myxomatosis. J. Ecol. 48, 287-306.

Torre, V. 1975. Synchonization of nonlinear biochemical oscillators coupled by diffusion. Cybernetics 17, 137-144.

Tramer, E. J. 1975. The regulation of plant species diversity on an early successional old-field. Ecology 56, 905-914.

Turing, A. 1952. The chemical basis of morphogenesis. Phil. Trans. Roy. Soc. Lond. B237, 37-72.

Vandermeer, J. H. 1975. A graphical model of insect seed predation. Amer. Natur. 109, 147-160.

Van Valen, L. 1974. Predation and species diversity. J. Theoret. Biol. 44, 19-21.

Viereck, L. A. 1966. Plant succession and soil development on gravel outwash of the Muldrow Glacier, Alaska. Ecol. Monogr. 36, 181-199.

Vivrette, N. J., and C. H. Muller. 1977. Mechanism of invasion and dominance of coastal grassland by <u>Mesembryanthemum</u> <u>crystallinum</u>. Ecol. Monogr. 47, 301-318.

Watt, A. S. 1947. Pattern and process in the plant community. J. Ecol. 35, 1-22.

Watt, A. S. 1955. Bracken versus heather, a study in plant sociology. J. Ecol. 43, 490-506.

Watt, A. S. 1957. The effects of excluding rabbits from a grassland B (Mesobrometum) in Breckland. J. Ecol. 45, 861-878.

Watt, A. S. 1960. The effects of excluding rabbits from acidiphilous grassland in Breckland. J. Ecol. 48, 601-604.

Webb, L. J., J. G. Tracey, and W. T. Williams. 1972. Regeneration and pattern in the subtropical rain forest. J. Ecol. 60, 675-695.

Went, F. W. 1970. Plants and the chemical environment. <u>in</u> E. Sondheimer and J. B. Simeone (eds.), Chemical ecology. Academic Press, New York.

Whittaker, R. H. 1962. Classification of natural communities. Bot. Rev. 28, 1-239.

Whittaker, R. H. 1965. Dominance and diversity in land plant communities. Science 147, 250-260.

Whittaker, R. H. 1967. Gradient analysis of vegetation. Biol. Rev. 49, 207-264.

Whittaker, R. H. 1970. Communities and ecosystems. Macmillan, New York.

Whittaker, R. H. 1972. Evolution and measurement of species diversity. Taxon 21, 213-251.

Whittaker, R. H., and P. P. Feeny. 1971. Allelochemics: chemical interactions between species. Science 171, 757-770.

Whittaker, R. H., and S. A. Levin. 1975. Niche: Theory and application. Dowden, Hutchinson and Ross, Inc., Stroudsburg, Pennsylvania.

Whittaker, R. H., and S. A. Levin. 1977. The role of mosaic phenomena in natural communities. Theoret. Pop. Biol. 12, 117-139.

Whittaker, R. H., S. A. Levin, and R. B. Root. 1973. Niche, Habitat, and Ecotope. Amer. Natur. 107, 321-338.

Williams, A. B. 1936. The composition and dynamics of a beech-maple climax community. Ecol. Monogr. 6, 318-408.

Williams, C. B. 1964. Patterns in the balance of nature. Academic Press, New York.

Williamson, G. B. 1975. Pattern and seral composition in an old-growth beech-maple forest. Ecology 56, 727-731.

Woods, F. W. 1960. Biological antagonisms due to phytotoxic root exudates. Bot. Rev. 26, 546-569.

Yarranton, G. A., and W. G. E. Green. 1966. The distributional pattern of crustose lichens on limestone cliffs at Rattlesnake Point, Ontario. Bryologist 69, 450-461.

Yodzis, P. 1976a. The effects of harvesting on competitive systems. Bull. Math. Biol. 38, 97-109.

Yodzis, P. 1976b. Species richness and stability of space-limited communities. Nature 264, 540-541.

Yodzis, P. 1977. Harvesting and limiting similarity. Amer. Natur. 111, 833-843.

Yodzis, P. 1978. Environmental fluctuations and the tenacity of equilibria. J. Theoret. Biol., in press.

Zedler, J. B., and P. H. Zedler. 1969. Association of species and their relationship to microtopography within old fields. Ecology 50, 432-442.

AUTHOR INDEX

Poore, M. E. D. 115

Porter, J. W. 10, 16, 42, 64, 102

Potts, D. C. 78

Potzger, E. 5, 70

Potzger, J. E. 5, 70

Price, P. W. 3

Prigogine, I. 11

Quarterman, E. 69

Randall, J. E. 80

Recher, H. F. 110

Reiners, W. A. 68

Rice, E. L. 42, 64, 65

Richards, P. 5

Richards, P. W. 114, 115‐

Risser, P. G. 64, 65

Rollet, B. 115

Root, R. B. 8

Rosenzweig, M. L. 111

Roth, R. R. 110

Roughgarden, J. 84, 98, 109

Ruelle, D. 50

Rützler, K. 4

Sagar, G. R. 9

Saila, S. B. 84

Sale, P. F. 4, 54, 71, 78, 110

Sansone, G. 160, 161

Sarà, M. 4

Schaeffer, R. 53

Schjelderup-Ebbe, T. 39

Schulz, J. P. 115

Scott, J. T. 69

Scriven, L. E. 11

Seigler, D. 3

Shafi, M. I. 64, 69

Simpson, E. H. 62

Sims, H. P. 2

Sinclair, A. R. E. 76

Singh, J. S. 64, 65

Sinkúle, J. 11

Skellam, J. G. 11

Slatkin, M. 17, 84, 102

Slobodkin, L. B. 13, 19, 76, 84, 102

Smith, A. P. 9

Smith, F. 76

Smith, C. L. 4, 110

Smoller, J. 17

Sparling, J. H. 9

Summerhayes, V. S. 100

Sutherland, J. P. 53

Swain, A. M. 5

Takens, F. 50

Tansley, A. G. 100

Taylor, D. L. 69

Terborgh, J. 10

Thomas, A. S. 100

Thompson, H. 2

SUBJECT INDEX

An underlined number indicates the page on which the corresponding
term is defined.

Bio— mathematics

Managing Editors: K. Krickeberg, S. A. Levin

Editorial Board: H. J. Bremermann, J. Cowan,
W. M. Hirsch, S. Karlin, J. Keller, R. C. Lewontin,
R. M. May, J. Neyman, S. I. Rubinow, M. Schreiber,
L. A. Segel

Volume 1:
Mathematical Topics in Population Genetics
Edited by K. Kojima
1970. 55 figures. IX, 400 pages
ISBN 3-540-05054-X

"...It is far and away the most solid product I have
ever seen labelled biomathematics."
American Scientist

Volume 2: E. Batschelet
Introduction to Mathematics for Life Scientists
2nd edition. 1975. 227 figures. XV, 643 pages
ISBN 3-540-07293-4

"A sincere attempt to relate basic mathematics to the
needs of the student of life sciences."
Mathematics Teacher

M. Iosifescu, P. Tăutu
**Stochastic Processes and Applications in Biology
and Medicine**

Volume 3
Part 1: **Theory**
1973. 331 pages.
ISBN 3-540-06270-X

Volume 4
Part 2: **Models**
1973. 337 pages
ISBN 3-540-06271-8

Distributions Rights for the Socialist Countries:
Romlibri, Bucharest

"... the two-volume set, with its very extensive bibliography, is a survey of recent work as well as a textbook. It is highly recommended by the reviewer."
American Scientist

Volume 5: A. Jacquard
The Genetic Structure of Populations
Translated by B. Charlesworth, D. Charlesworth
1974. 92 figures. XVIII, 569 pages
ISBN 3-540-06329-3

"...should take its place as a major reference work.."
Science

Volume 6: D. Smith, N. Keyfitz
Mathematical Demography
Selected Papers
1977. 31 figures. XI, 515 pages
ISBN 3-540-07899-1

This collection of readings brings together the major
historical contributions that form the base of current
population mathematics tracing the development of
the field from the early explorations of Graunt and
Halley in the seventeenth century to Lotka and his
successors in the twentieth. The volume includes
55 articles and excerpts with introductory histories
and mathematical notes by the editors.

Volume 7: E. R. Lewis
Network Models in Population Biology
1977. 187 figures. XII, 402 pages
ISBN 3-540-08214-X

Directed toward biologists who are looking for an
introduction to biologically motivated systems
theory, this book provides a simple, heuristic
approach to quantitative and theoretical population
biology.

Springer-Verlag
Berlin
Heidelberg
New York

A Springer Journal

Journal of

Mathematical Biology

Ecology and Population Biology
Epidemiology
Immunology
Neurobiology
Physiology
Artificial Intelligence
Developmental Biology
Chemical Kinetics

Edited by H.J. Bremermann, Berkeley, CA; F.A. Dodge, Yorktown Heights, NY; K.P. Hadeler, Tübingen; S.A. Levin, Ithaca, NY; D. Varjú, Tübingen.

Advisory Board: M.A. Arbib, Amherst, MA; E. Batschelet, Zürich; W. Bühler, Mainz; B.D. Coleman, Pittsburgh, PA; K. Dietz, Tübingen; W. Fleming, Providence, RI; D. Glaser, Berkeley, CA; N.S. Goel, Binghamton, NY; J.N.R. Grainger, Dublin; F. Heinmets, Natick, MA; H. Holzer, Freiburg i. Br.; W. Jäger, Heidelberg; K. Jänich, Regensburg; S. Karlin, Rehovot/Stanford CA; S. Kauffman, Philadelphia, PA; D.G. Kendall, Cambridge; N. Keyfitz, Cambridge, MA; B. Khodorov, Moscow; E.R. Lewis, Berkeley, CA; D. Ludwig, Vancouver; H. Mel, Berkeley, CA; H. Mohr, Freiburg i. Br.; E.W. Montroll, Rochester, NY; A. Oaten, Santa Barbara, CA; G.M. Odell, Troy, NY; G. Oster, Berkeley, CA; A.S. Perelson, Los Alamos, NM; T. Poggio, Tübingen; K.H. Pribram, Stanford, CA; S.I. Rubinow, New York, NY; W.v. Seelen, Mainz; L.A. Segel, Rehovot; W. Seyffert, Tübingen; H. Spekreijse, Amsterdam; R.B. Stein, Edmonton; R. Thom, Bures-sur-Yvette; Jun-ichi Toyoda, Tokyo; J.J. Tyson, Blacksbough, VA; J. Vandermeer, Ann Arbor, MI.

Springer-Verlag
Berlin
Heidelberg
New York

Journal of Mathematical Biology publishes papers in which mathematics leads to a better understanding of biological phenomena, mathematical papers inspired by biological research and papers which yield new experimental data bearing on mathematical models. The scope is broad, both mathematically and biologically and extends to relevant interfaces with medicine, chemistry, physics and sociology. The editors aim to reach an audience of both mathematicians and biologists.